Escaping Normal

Drifting
Souls

Marella Sands

Word Posse

Dedication
To Those Who Told Me Their Stories

Previous Escaping Normal Titles
Descending Skies
Visiting Strangers

The Angels' Share Books
Volume 1: Through a Keyhole, Darkly
Volume 2: What the Thunder Said
Volume 3: The Chair She Sat In
Volume 4: With Sleepless Eye
Volume 5: Past the Isle of Dogs
Volume 6: When We Had Feathers

Other Word Posse Books by Marella Sands
Pandora's Mirror
Fortune's Daughter
Restless Bones

Ring of Fire Press Books by Marella Sands
Perdition
Purgatory
Perfection

Catch up with the author on various platforms:
Facebook: facebook.com/MarellaSands
YouTube: Search *Marella Sands* once there!

Note: photos printed in this book will be at 300 dpi *at best*. Look on my website for higher res pictures.

ISBN-13: 978-1-944089-27-6

Table of Contents

Where to Begin?

We're all ghosts. We all carry, inside us, people who came before us.
Liam Callanan, *The Cloud Atlas*

I shall not commit the fashionable stupidity of regarding everything I cannot explain as a fraud.
Carl Jung

In one aspect, yes, I believe in ghosts, but we create them. We haunt ourselves.
Laurie Halse Anderson, *Wintergirls*

*
**

The obvious question is, *what is a ghost?* People use the word as if we all agree on what a ghost is, but do we? I once asked my students to consider the question, but they seemed incapable of actually thinking about it terribly deeply. "It's a human soul," said one, leading me to ask the clear follow-up, "Okay, so what's a soul?" Silence.

Another student said, "It's energy." I asked in return, "What kind of energy?" The student was stumped. Eventually, he could only repeat, "Just...energy."

It's easy to talk about ghosts, but certainly less easy to decide what a ghost actually is. For that reason, I'll be writing several books dividing the things that may or may not be lumped under the generic term "ghost" into several categories.

Ghosts, for the purposes of this series, will be considered to be some kind of continuation of a human post-death. Other forms of entities that I will focus on in future books will include shadow people, poltergeists, angels, black-eyed children, revenants, and demons.

I have had no experiences with ghosts myself. Now, when I was about three, I thought I saw five of them coming for me in the night. Nowadays, I'm pretty sure they were merely the lights from a passing car moving across the wall. But I remain fascinated by the idea that ghosts may be out there, perhaps all around us, and we are simply not able to sense them. Is becoming a ghost a fate we can all look forward to (or dread, as the case may be)?

In this book, we'll look at the history of ghosts, at least a few of the kinds of ghosts people have documented and believed in, and what might explain the phenomenon. We'll also talk about various folklore ghosts, some of whom are weak and not scary at all, while others become powerful and must be appeased if they are not to do harm to the living. Ghosts have been consulted as seers and sought after as tourist attractions. They have been pitied, loved, venerated, and feared.

For many, the mere fact that ghosts provide evidence of an afterlife is the most important thing. Death is one of the constants of our lives that we have a tendency to avoid thinking about (at least in Western cultures). In one of my university courses, we spend a class period discussing issues surrounding death, how our culture copes (or doesn't), and what students think about the customs and traditions of their own culture. Would they want their remains treated in whatever their culture considers traditional? Would they prefer some other way to rejoin their physical body with nature? Most students are able to participate in the discussions on that day fairly well, but others do not, and make sure I hear about it on the student evaluation form when the semester's over. One semester a student complained that discussing death in class was "inappropriate."

Many people are uneasy around death, anything to do with the subject, and certainly are creeped out by the bodies of those who have died. Or, euphemistically, *passed on*. Or *crossed the divide*. Or, as the ancient Egyptians would say, *gone West*. Taboos are strung like tripwires throughout every culture on the planet. Can you touch a dead body? Should you? Do you bury your loved one? Cremate them? What ceremonies do you hold to convince their spirit to move on? What might the dead do if you fail in any of your duties?

Is handling all this a family affair? Something the entire community gathers to take care of? Or are there specialists to deal with the unpleasantries surrounding death and the afterlife?

These questions have been answered in many ways over the millennia, and are outside the scope of these books because the disposal of the body itself doesn't rate as paranormal. That's just housekeeping.

The question is, what happens to *us* once we've crossed that divide, assuming anything happens at all. But ghosts, and the stories surrounding them, give us answers that, perhaps, unnerve us, but ultimately are designed to make us feel better. Yes, death happens, but it's not the end of us. Ultimately, we believe we have evidence that something in us is not extinguished by death.

And that, perhaps is the most important thing there is to say about ghosts. As much as they might frighten or disturb us, they give us hope.

> If you are at all knowledgeable about ghosts, you know this book barely scratches the surface of the topic! Let me hear from you about your experiences or what you'd like to see covered in a future installment of *Escaping Normal.*

The Toxic Dead

The terms we use for what is considered supernatural are woefully inadequate. Beyond such terms as ghost, specter, poltergeist, angel, devil, or spirit, might there not be something more our purposeful blindness has prevented us from understanding? We accept the fact that there may be other worlds out in space, but might there not be other worlds here? Other worlds, in other dimensions, coexistent with this? If there are other worlds parallel to ours, are all the doors closed? Or does one, here or there, stand ajar?
Louis L'Amour, *The Haunted Mesa*

If ghosts are real, death is not.
A.J. West, *The Spirit Engineer*

*
**

The scene: northern Spain, approximately 400,000 years ago. A group of early humans descend into a cave with the body of one of their own. Their destination: a deep shaft in the cave down which they lower the body of their deceased fellow. We'll never know with 100% surety, of course, what they hoped to accomplish by this disposition. Did they mourn? Pray? Were they afraid while they stumbled through the cave and braved the crazy leaping shadows created by the rock walls and the flickering light of their torches? What kind of terror or hope did that lightless shaft down which they placed their dead invoke in them? How relieved were they, at the end of it all, to return to the above-ground world of trees and streams and clouds? Of night and day, winter and summer?

All we know for now is these early humans, who were previously known as "archaic" *Homo sapiens*, but who are now classified as their own species of *Homo heidelbergensis*, did something in the cave of Sima de los Huesos, near Atapuerca, Spain with some of their dead. Due to the fact that the dead found in the shaft do not represent a cross-section of a typical community (the very young and very old do not seem to have been taken to the cave shaft, for instance), the current theory is that those placed in the cave represent a particular group, though of course how that particular group was defined is beyond our ability to guess. We simply know that the people in the shaft were people whose fates after death were different than others in their community.

Also, the fact that small bones like the hyoid are found in the shaft indicate that the dead were placed there while still an intact corpse. Because there are no signs of occupation in the cave, this means that the living members of the community needed to carry their dead from wherever they lived to the cave. They then had to face the stygian darkness while carrying torches and a dead person. They had to stumble and crawl through this frightening underworld until they reached the shaft where they deposited their deceased community member. They had to realize that leaving a body in the shaft made it irretrievable. What the shaft swallowed up it would never return.

The activities at Sima de los Huesos are the earliest we have found, so far, of death rituals being carried out by members of the genus *Homo*. But what did it mean to those people to leave their dead there? Did they have much of a concept of an afterlife? Of magic? Or of a soul? They must have had some kind of nascent beliefs or they wouldn't have gone to the trouble. *Something*, after all, made them carry a dead person for who knows how far, caused them to brave the eternal night of the cave, and to leave the person behind. And *something* had marked those individuals as requiring or deserving a different treatment after death than the rest of their group.

To take it one step even further, *something* made these people think that the dead needed some kind of post-death disposition at all. Our closest relatives, chimpanzees, may keep vigil over a body for a short while, perhaps until they realize that individual will not be moving on with them, but then they leave the remains without performing any kind of parting rituals. The deceased chimp is abandoned much like any other inanimate object.

We can assume that early humans did the same. But, at some point, this changed. Early humans began realizing that the corpse was in a liminal state of being both a recognizable person and an inanimate object at the same time. Simply walking away no longer became a viable option.

Timothy Taylor, in his book *The Buried Soul*, feels that the original reason for rituals conducted after the death of a community member, were to deal with what he refers to as *the toxic dead.* Not toxic as in poisonous, but toxic as in occupying a terrifying middle space that is neither one thing nor another. Whether it's pregnancy and childbirth, adolescence, marriage, joining a fraternal order or the military, or death, the transition from one social status to another (woman to mother, civilian to military, child to adult, living person to dead) is something the community hems in with rituals designed to control and demarcate the stages of change. Being in a liminal state is both powerful and frightening. The community must do something to cope.

> **Fun Fact**
> In the Twi language of West Africa, the word *nwunu* is used for anything that is cold that is not a person, such as inanimate objects. Cold people are *awu.* However, once someone dies, the word *nwunu* applies.

It appears our ancestors and hominid cousins may have been doing this for close to half a million years.

Flower Children?

However, we have to be cautious of a trap that is notoriously hard to spot, and that is to make assumptions based on our own biases. Simply referring to something as a *burial* reveals something about ourselves and the lens through which we interpret things symbolically, which may or may not actually apply to situations in the past.

One example of this is the alleged burials in Shanidar, a cave in modern-day Iraq. People were fascinated and shocked when the original researchers declared that they now had proof that Neanderthals buried their dead. They even put flowers on the graves, just like we do!

I recall reading an article about Shanidar when I was very small. I, too, was blown away by the ideas presented. I wasn't the only one. Jean Auel clearly used the Shanidar people as a source of inspiration for her Neanderthals in her book *The Clan of the Cave Bear*.

Considering that the work at Shanidar was happening while the world was in the midst of the societal upheavals of the 1960s, it shouldn't, perhaps, surprise us that the researchers were prone to describe the Neanderthals of Shanidar as a peaceful people who took care of their dead, tended the disabled, and put flowers on the graves of their loved ones. The researchers waxed a tad bit poetic, describing Neanderthals as having a sense of spirituality heretofore not considered by modern-day archaeologists. Neanderthals were the original "flower children," as it were.

It didn't take long for others to look at the data and come away with different conclusions. The "graves" might have been individuals caught in rock falls. The pollen might have been blown in by the wind or brought in on the fur of wild animals who sheltered in the cave during bad weather. Since the "graves" lacked grave goods, it was difficult for some to see these as deliberate burials. Rock falls are not uncommon in caves, and Shanidar is prone to them, so the idea that some individuals were occasionally caught by them isn't even a stretch of the imagination.

Not even those ideas could escape the trap: while the supporters of the theory of peaceful Neanderthals who lovingly buried their dead might have been influenced by the cultural tropes of the 1960s, the supporters of the theory of rock fall-killed Neanderthals may have been influenced by old ideas of Neanderthals as brutish and stupid, unable to speak, and incapable of any sense of spirituality.

Timothy Taylor looked at the data and came up with another take on it, one which follows his theory of death rituals being mainly concerned with isolating the toxic dead from the living. In his analysis, the fact that the Shanidar "burials" do not represent a cross-section of society, but only encompass those on the margins (the very old, the very young, the disabled) shows that those people weren't merely victims of random rock falls, but deliberately chosen to be placed inside the cave in order to isolate them from both the living and the rest of the dead.

To Taylor, what Shanidar suggests, is that most people in Neanderthal society faced a different afterlife than those in the cave. Perhaps it was something similar to the life they'd led as living people. Who knows? But those placed in the cave were being taken to a liminal place where there is no change of day to night, no seasons, no fluctuations in temperature. In other words, a timeless place where the normal rules of the world are turned upside down. Far from being a location where loved ones were placed for their incorporation into the land of the dead, Shanidar was a place representing the abandonment and isolation of those dead from then rest of their people for all time.

If that seems extreme or even cruel, well, yes, that's part of Taylor's point. Humans can often be cruel to each other, and that doesn't necessarily stop just because some of those humans are dead.

Why Consider Sima de los Huesos and Shanidar Here?

We can't really say what those *H. heidelbergensis* individuals thought or believed when they dropped certain members of their community into a night-dark shaft deep inside a cave. We don't really know if the Neanderthals of Shanidar gave much thought to their dead, or whether they differentiated between those dead that they treated well, and those who were barred from the afterlife everyone else got to inhabit. The important part is, it seems we are looking at certain times in the distant past where humans and their antecedents were beginning to consider that something happened to them after death. Not just to their bodies, but to what we would term today as a personality, or a psyche, or a soul. *Something* animated people while alive, a something that was no longer there after death. In order to cope with this something (I'll call it a soul, though clearly that shows my own cultural bias), rituals had to be developed to delineate what was living, what was dead, and even what was in between.

I'm not saying that these cases indicate that the people of the distant past believed in souls per se. But they obviously were beginning to give thought to that age-old question, *what happens to me when I die?* Not just to my body, but to *me*. Corollary questions then arose: if people have souls, where do those souls go? Do they all go to the same place? Could they hang around the campfire and the people they miss and cause trouble? Are they watching us now? How do we make sure we keep the lines between the living and the dead distinct? What can go wrong if we don't?

In other words, people were beginning to conceive of the idea of ghosts.

Leveling Up...Or Not

We tell stories of the dead as a way of making a sense of the living. More than just simple urban legends and campfire tales, ghost stories reveal the contours of our anxieties, the nature of our collective fears and desires, the things we can't talk about in any other way. The past we're most afraid to speak aloud of in the bright light of day is the same past that tends to linger in the ghost stories we whisper in the dark.
Colin Dickey, *Ghostland: An American History in Haunted Places*

It occurred to me that if I were a ghost, this ambiance was what I'd miss most: the ordinary, day-to-day bustle of the living. Ghosts long, I'm sure, for the stupidest, most unremarkable things.
Banana Yoshimoto, *The Lake*

*
**

In the *Odyssey,* when Odysseus is told by Circe to go to the land of the dead to consult the late seer Tiresias, he discovers the dead are both frightening and pitiful. What they are not, is powerful.

In order to contact the dead, Odysseus must perform certain rituals, including sacrificing animals, in order for their blood to be available for the dead to drink. Without this imbibing of blood, the dead would not be strong enough to communicate with the living.

While Odysseus is filled with dread at the thought of being confronted by the dead, he does as he is told. Soon enough, shades of the dead come forward to drink the blood. One of them, Elpenor, is the soul of one of Odysseus' men who was quite alive the last time Odysseus saw him mere hours before. Therefore, he is surprised to see Elpenor in front of him. Does Elpenor do anything particularly scary? No, he simply requests that his body be given a proper burial.

Odysseus is also surprised by the appearance of his mother, as he was unaware of her death. She drinks the blood and is able to tell her son that his father is still alive. Odysseus attempts to embrace her, but she is merely a shade. He is unable to touch her.

The seer Tiresias approaches and gives Odysseus some useful information such as why Poseidon is angry with him, and what will happen to him on his way home provided he does or does not do certain things. This is the information Circe sent him to get, so this whole scheme seems to have worked.

Before he can go, Odysseus is approached by old comrades such as Agamemnon and Achilles. Mostly, they use what little time they have with the living to whine about how they died and how much they don't like their current state. Achilles says he'd rather be a living slave than be the king of all the dead.

It becomes clear to the reader that the Homeric dead are powerless. Even if they are angry, their after-death grumpiness affects no one but themselves. Sarpedon, who was killed during the Trojan War by Patroklos, is still angry about his death, but it's not as if his fellow dead, or the living, are influenced by this at all. His bad temper may be making his personal afterlife more miserable than it might be otherwise, but that's as far as it goes.

Despite the fact that listening to the departed bitch about their fate shouldn't be that frightening, Odysseus is affected enough that he retreats as quickly as he can. Perhaps this is due less to a fear of the dead themselves than the fear of his own mortality. After all, one day, this, too, will be Odysseus' fate: to become a substanceless shadow with no strength, and nothing to do but complain.

Things hadn't gotten much better for the dead by the time of the Classical Greeks. While the murdered dead of Classical times may have been able to petition the gods to take vengeance on their murderers, they are not able to do so directly. They cannot act, so the gods must act for them.

The Greeks appear to have been attracted to the idea of consulting the dead, but there seems to be little evidence that this ever did much good. The dead were not merely powerless and witless, they were often uninformed of what was going on in the world. Therefore, even after going to the trouble to contact them, the living found that the advice of the dead had a tendency to be either useless or merely obvious. Sometimes, the dead's only concern was to have someone tell their loved ones that they still loved them.

For the most part, the dead of ancient Greece were simply pitiful. But in other cultures, the dead could wield, or threaten to wield, a great deal of power.

Theater of Transgression

Let's say you've grown up in a society where, because you were born, say, with a withered leg, you've been considered someone the gods have marked as their own. You're a person with challenges, sure, but in return, you've been granted special access to the supernatural. Perhaps you are believed to be able to predict the future, or ensure the fertility of the land and people. But lately, your predictions aren't panning out. Or perhaps a disease has ravaged the community's cattle and you were as surprised as anyone.

You're a bit worried, but perhaps not too much. The gods have given you their blessing, so surely any ill-luck is temporary. You're the gods' chosen mouthpiece, after all. But one day you see the mob coming for you, and you know your fate is sealed. Well, they might kill your body, but you're the special one that the gods love, right? They'll help you get your revenge from beyond the grave.

Except you realize, when the mob gets closer, that they're very aware of your supernatural abilities, and know exactly how to counteract them. You no longer have any power in this world, and now you know you'll have none in the next.

This is the scenario Timothy Taylor considers may have been the fate of those whose bodies have been found in northern Europe's bogs ("bog bodies"). Taylor looked at the demographics and physical health of those from the bogs, and noticed that those who would have been nearly powerless in life (like babies, and those whose bodies show signs of manual labor and so may have been servants or slaves) seem to have been simply killed and tossed in the bog, while those who appear to have been of more elite status suffered horrendous deaths. Tollund Man, for instance:

- was poisoned;
- was struck in the back;
- was hit over the head;
- was strangled;
- had his throat cut;
- was drowned.

Any one of these methods would have been enough if the goal were simply to kill his body. But clearly, by making his death as violent and brutal as possible, those who killed him were going for more than mere death. They wanted something else entirely.

They wanted to kill, or imprison, the soul.

Taylor posits this because these bodies were placed somewhere they would not decay, and that, no matter the violence against the rest of their bodies, their faces were not marred. The preserved body, and clearly recognizable face, would keep the soul tied to the corpse itself, or at least to its immediate environs, making any revenge from the afterlife impossible, no matter how powerful the person had been in life.

But what could have prompted the people of the community to do this to one of their own? After all, they weren't simply killing a killer, and assuming the gods would punish the person in the afterlife. They were killing someone *and forcing both body and soul to remain in the bog forever.*

This is where the idea of the *theater of transgression* comes in. In societies where life is precarious, and being like everyone else is greatly valued because it helps the community survive, those who are different can easily become singled out. Born with albinism, or teeth? What if you're someone with mental or physical challenges? What if you have odd birthmarks? What if you seem incapable of accepting a way of life everyone else feels is the proper way to live? You may be too different for those around you to accept.

Heck, what if you're an identical twin in a world where it's understood that people are distinct individuals? Identical twins violate that understanding.

If any of these (or any other difference) applies to you, well, you may be committing a crime against your society. In a land of round pegs going in round holes, the square peg is likely to be censured in some way. Or removed from society. Or even, in extreme cases, cut out of life and afterlife entirely.

The reason this follows our conversation on the power of ghosts is that, if the dead had no power, no one would bother killing (or trapping) a soul. Consequently, the dead, at least the dead of the elite, must have been capable of wielding great power. Rather than face that power, society imprisoned it, and made the living safe from the dead.

From Terror to Curiosity

For most of human history, ensuring the safety of those left alive from those who have departed their physical bodies, has been a preoccupation of those left behind, possibly even to the point of obsession. I doubt it's easy for many of us these days to truly understand the sheer terror the dead could invoke, just as it's difficult for us to truly understand the terror of the dark that people in past centuries felt. A couple of years ago, while camping, my husband and I saw quite a few pair of glowing yellow eyes in the forest around us. Yes, they were raccoons. No, we weren't really frightened. But the experience was just creepy enough to give me a tiny glimpse into the mind of someone who did not know what lay beyond the limit of their fire's light.

However, over the past few centuries, while some fear has remained, the outright terror that must have gripped many people in the past, has, to some extent, faded away. Instead, people replaced fear with curiosity.

Was it enough to know that grandma had died and moved on? What if we could still speak with her? Could that even be possible?

What form had grandma taken? Was she a spectral mist? A humanoid form in a bedsheet? Did she moan at night? Does she know what's going on with her descendants? Is she angry that the family sold the farm? What, for that matter, is she *doing* with her afterlife?

Perhaps it was the Age of Reason that began changing things. Instead of fearing death and the afterlife, and the power of the dead, people began bringing the unknown out of the darkness (so to speak) and into the light of the Scientific Revolution.

It was time for an overhaul of the spiritual world from the earthly side of things.

Things That Go Bump Up the Stairs

Now I know what a ghost is. Unfinished business, that's what.
Salman Rushdie, *The Satanic Verses*

The people you love become ghosts inside of you, and like this you keep them alive.
Rob Montgomery, poet

*
**

You sense something odd in your house. You hear the sounds of footsteps coming down the stairs, but they always stop at the landing. Sometimes, you smell roses.

Do you have a ghost in your house? Or is it a spirit? Or is there a difference?

Many people use words like *ghost* and *spirit* interchangeably. However, some people split them up into two different categories, with a *ghost* being the spectral remains of a human being that is trapped in a particular place, such as your house. These ghosts come in several varieties.

A *spirit*, on the other hand, is a human soul that is not tied to any particular person or place. It can travel throughout this plane of existence, and may even be able to cross dimensional boundaries. This is opposed to things referred to as *demons* or *shadow people*, which are other classes of supernatural entities that were never human.

Whether or not this distinction is actually useful, I don't know. Let's look at some of the varieties of these spectral entities and see if we can figure that out.

Only Leftovers

Perhaps one of the most common types of hauntings that is reported is a *residual haunting*. This is a type of haunting where a sound or smell or apparition does the same thing repeatedly. It has no mind; it is simply an echo of the past being replayed in the present. So this barely qualifies as a *ghost*, since, while it does seem to be the "remains" of a human, it is not anything with mind or will. It is simply a memory.

The explanation for residual hauntings is referred to as the Stone Tape Theory, after a BBC production of a ghost story called *The Stone Tape*, which aired in 1972. The idea was proposed several times before the 1970s, perhaps first by Charles Babbage in 1837, though his theory dealt exclusively with sounds leaving an impression in the air. While no one has yet proposed a mechanism by which these hauntings might occur, the idea that strong emotions might somehow imprint onto rocks, other objects, houses, or even forests, has proven popular, at least in a few countries such as Great Britain and the United States.

Because the more modern form of this idea of residual hauntings can be traced to a particular time and even one particular television show, skeptics point to it as a strong indication, if not actual proof, that the paranormal does not exist. After all, if residual hauntings were an actual thing, why would they suddenly have become a popular form of haunting only since the 1970s? But that skepticism hasn't kept people from adopting residual hauntings as a favorite explanation for paranormal phenomena. In fact, it is difficult now to find any ghost hunting book or article where residual hauntings are not presented as fact.

I have personally heard of a local residual haunting where the residents of the historic building sometimes hear someone walk up the staircase. But the sounds never go farther: no footsteps are ever heard walking the hallways or even coming back down the stairs! Residual hauntings tend to be very specific in nature (like, say, footsteps) and limited (going up the stairs). And just like a recording on magnetic tape, they have a tendency to fade after time. If at first people notice the footsteps on the stairs a few times a week, they may, after a few years, realize they are only hearing the noise once a week, or only a few times a year, and the sounds have gotten much fainter. What was one audible throughout the house becomes something that one can barely hear, and only when the house is completely quiet. Eventually, the footsteps (or whatever form the residual haunting takes) disappear entirely. The tape has run itself out.

Taking Care of Business
Some ghosts linger due to wanting to care for the things they've left behind. The classic "caretaker" ghost is one who used to live in a residence and becomes unhappy when something is changed. More than one paranormal researcher has pointed out that many of the hauntings they are called to investigate begin shortly after someone makes a repair or an update to a building.

Even worse is when someone has promised a previous resident that they wouldn't make changes, and then, once that resident has passed on, either changes their mind, or discovers problems with the structure that requires repairs. Once that happens, the current resident will claim to experience the loss of items, strange sounds, or even apparitions. The resident will assign responsibility of these incidents to a caretaker ghost who wants to express their disapproval for the renovation project.

One story related to me was that of a woman who had moved into a house that she was told was haunted by a boat captain. She heard the odd noise sometimes, and spotted a shadow of someone on the other side of a door at some point in time, but overall, her experiences were rare and not bothersome. She felt like, if there really were a ghost in her house, that it was not malevolent. It was simply sharing the home with her without interfering in her life. Then came a day where she was climbing a ladder and lost her balance. As she began to fall backward, she felt a strong hand press into her back and nudge her to safety. She was able to keep from falling and was saved from injury or even death.

Stories like this show that the kind of ghost that hangs around the places with which they're familiar, and who are aware of the "changing of the guard" (so to speak), can be helpful and supportive if they wish. Or, if they don't wish, they can become disgruntled, or even angry, if they perceive that the things they care about are not being respected. Far from moving on to an afterlife, they remain to watch over the things they cared about in life.

I Spy With My Little Eye?

Another form of ghost is sometimes referred to simply as *the deceased loved one*. Many stories of these phantoms are quick to emphasize that the people who saw the deceased do not realize the person is deceased when at the time they see them.

One story I've read involved a servant in a large country household in England during World War I. A servant looked up from her work and saw the son of the landowner standing across the pond that was on the property. She was so excited that the young man had returned home from the war that she ran to tell everyone about his arrival. A sudden bustling of activity was produced as cooks rushed to make a special dinner and the young man's room was prepared for his arrival. The family waited eagerly for the man to finish his walk up the drive and come into the house, home from the war at last.

But the man did not come. His tea grew cold. The family began to become worried. The servants looked about the house and grounds. But no trace of the young man could be detected. Eventually, everyone assumed the servant who had spotted him had been mistaken and the disappointed household went back to waiting for the young man to return from the war.

As you may have already guessed, the family shortly thereafter received a telegram telling of the young man's death in battle, at precisely the day and time he was spotted by the servant.

Whether or not this particular story is true, or is just an urban (countryside?) legend, isn't really important. It is indicative of the kind of stories still told to this day. Someone sees a person they know, but the person can't be found again a moment later, and sometime after that, the person is told that the individual they saw couldn't have been physically present because they had died at that very time somewhere else.

A similar kind of story is one where the individual spots someone they *don't* know, and the person disappears almost as soon as they are spotted. So instead of *the deceased loved* one, we're confronted by *a deceased stranger*. A person gets up in the night to get a drink of water, spots a stranger in the house, and in a sudden fright that they've seen a burglar, checks again to make sure of what they saw. But the nighttime apparition has already disappeared. No burglar can be found, and police, assuming they are called, will find no evidence of a break-in. The mysterious visitor was there and gone, and yet is not anyone the witness can identify. The apparition came with apparently no reason or connection to the person they appeared to.

Sometimes, this kind of apparition seems attached to a place. In one story, a woman and her child were visiting a friend. The woman's child spotted a crying child on the stairs and rushed to her mother to tell her about the other child. The mother, of course, was upset that a strange child might have gotten into the house, and goes to investigate, but no crying child can be found. The mother then goes to the homeowner to ask if there is another child in the house, because her child saw one crying on the steps, and the homeowner (as you could surely guess) grows pale, and recounts that her own child had seen a crying child on the steps decades before.

Sometimes, these apparitions may be more like residual hauntings, doing the same thing over and over, but sometimes they may speak to the witness or appear frightened and disappear once they notice they've been spotted. It's as if they are trapped somehow, and are aware they are trapped, but can do nothing about it.

G.N.M. Tyrrell, in his 1963 book *Apparitions*, laid out four kinds of ghosts. I include his list here because he mentions one form of ghost that is manufactured by the living, and I had not seen anyone do that before.

Tyrrell's list includes:

Living Ghosts: Sometimes, a person has an apparition of someone they know who is still alive. Since Tyrrell considers this form of ghost to be deliberately created by the person who is seen far from their physical body, perhaps this might best be classified as some sort of astral projection rather than a ghost, as least as we are using the term *ghost* here.

Crisis Apparitions: Like the deceased loved on, this is an apparition seen by others upon someone's death. The story of the World War I soldier fits this category perfectly.

Post-Mortem Hauntings: For some reason, Tyrrell put a time limit on this. A post-mortem apparition is seen within twelve hours of death, making it a sort of middle ground in between a crisis apparition and a more traditional haunting.

Hauntings: This category is the more standard haunting. The person is seen somewhere they were also seen in life, long after the twelve hour window for the Post-Mortem Haunting.

Most people in the Western world have probably heard stories of these particular kinds of spectral visitors throughout their lives, or have seen television shows or movies about caretakers, residual hauntings, and deceased loved ones (or strangers, as the case may be). I have collected a few from those who have answered my call to tell their own stories, and will recount them now.

Personal Story: Lauretta

It's an unexplainable story, let's put it that way. I would say I was pretty convinced most of my life that I had a ten-minute conversation with a dead person. I was staying with my cousin; I used to spend weeks at a time with them. We were playing with the kid next door and it was the first time I'd ever been in the house. My cousin Eugene and I had gone upstairs, and we a had a train or a ball or something. It rolled away from us. I went down the hallway and said, "I'll get it," and Eugene said okay, and he ran back downstairs.

I went into this room and there was this old lady sitting there in this rocking chair, and she was like, "What are you doing here?" And I said, "I'm playing with my cousin." We just started talking. I remember she had this old Victorian kind of lamp that had this really warm glow to it. She also had a cedar chest. I said, "You've got a cedar chest, just like my grandma." She opened it, and there's all this stuff in the cedar chest and she starts pulling out pictures. She said, "This is when I went to Paris." And she's showing me all these mementos of her life. I get really emotional thinking about it because it was just so real, and it was like this was this person's life that she was sharing with me for whatever reason.

After a few minutes, I said, "Well, I gotta go now." And she said, "Come back if you ever get the chance." I went downstairs and Eugene and I continued playing with the kid and then we left. We went back to my aunt and uncle's house and my aunt asked, "What'd you guys do over there?" Eugene was like, "We played." I said, "I had this conversation with their grandma."

My aunt got this really weird look on her face and the color drained and she said, "What do you mean, their grandma?" I said, "Well, there was this old lady upstairs in one of the bedrooms. She was really cool. She was showing me all these pictures and stuff." And my aunt, you could see she was just physically shaken by this. I asked, "What?" She said, "Don't worry about it."

So she calls over there and said, "My niece was just over there, and she said she was talking to some old lady in one of the rooms upstairs." The neighbor got really quiet and said, "There's nobody here." So my aunt called my mom, and my mom was freaking out. My aunt was really concerned.

I think the reason why my aunt had that reaction was because the neighbor had said to her that there were always weird things going on in the house, especially in one of the bedrooms upstairs. If they turned the light off, they'd go back and it'd be on. They'd shut the door; when they went back, it was open. They'd pull the blinds down in the window, and they'd go back and they'd be up. That kind of stuff.

The old lady was so calm and sweet, and the stories she was telling me about travel and all that stuck with me. I mean, I've never forgotten it. So that's my ghost story.

Personal Stories: Kate

I don't see ghosts, except for this one time. I got a picture of whatever, and it was the day that my friend passed away. She worked third shift and I had an on-call job at the hospital. And that was the only time we ever saw each other. She looked gray that day. She looked horrible. She said she hadn't slept for 24-36 hours and I said, "Let me do some Reiki on you."

And so I did my thing and you could see pinked up while I was doing my thing. Then we got to talking about weird stuff, and she talked about dying. I said, "Well, you know, if you ever do decide to die before me, come look me up. I might have the opportunity to know that you're there." Three months later, I'm randomly taking pictures in the dark, because I had insomnia a lot. And there's this pink light. I posted it on Facebook. And everyone to a person said, "Who is that?" They didn't say WHAT is that? They wanted to know WHO it was.

Kate's photo of the pink streak. Photo used by permission.

I went to work and found out this friend had passed away unexpectedly. It was supposed to be her first day back after knee surgery. She hadn't shown up for her shift; she'd had a cardiac event and was gone.

The cool part of that story is, later, I put out the thought that, if you have any message you want to give to anybody, let me know. I saw a light falling toward one of my crystals. I have a lot of crystals. The light fell toward the crystal three or four times. And I said, okay, so you want me to give a crystal to a family member. I got the impression she said yes. I thought, she's gonna want me to give that to her son. He is not gonna understand. And I don't want to do it.

All of a sudden it popped into my head, she wanted me to give it to her sister, whom she hadn't talked to in fifteen years. Never would have occurred to me. I said, "Really?" And the light fell toward the crystal and I'm like, "Okay." So, I gave the sister the crystal and she said, "Do you really think she wanted me to have this?" I said, "I don't know, but yeah."

I have another story about speaking with people who have passed on. The first time it happened, I was at my father-in-law's funeral. I had not really spoken to the man maybe more than six times in twenty-five years. He was Silent Joe. He just never talked. And so I'm sitting there and I'm really sad for my mother-in-law because he'd been her whole life. I'm just bawling. For her. And, all of a sudden, it was like something just hit me in the chest and it was like there was this feeling of this love that you can't even imagine. I can still feel it.

So then years later, Grandma dies, and we're at the cemetery, This was in January; it's minus fifteen degree weather and I'm just freezing. All of a sudden, a hand touched my back and instantly I was warm, didn't shiver one bit for the rest of the time.

So, another friend, she was a patient of mine, she knew she was dying. And I said, "If you pass away, just come and try and find me." I came in on Monday and she had passed. I went in the room and cried and cried and, all of a sudden, boom. The same feeling hit me. So, that's basically what I've had, you know, with people who've passed on.

I do have a funny story, though, that kinda has to do with this. My mother and I, we were having this conversation, and I said, "I read an article that said that after people pass away that they sometimes leave a scent for someone. When you pass away, what scent would you leave me to let me know you were there?" Mom, during her lifetime, was very gaseous. She said, "I don't know, how about a nice juicy fart?"

Two years later, she passed away. A couple of days later, I'm lying on the couch in the dark and I'm thinking about her and I'm crying and all of a sudden there's this horrible smell. I said, "Mom?" And I started laughing and crying, and I'm like, oh no, it's just toxic sludge in the basement coming up from the septic tank, because it was that bad. And once I thought that, that smell was gone.

Weird stuff has happened to me all my life.

Local High School Legend: Marella

This story was recounted to me by one of my high school teachers. It seems that, on their way to a party, four high school students were killed in a car crash. No one at the party was aware of this for some time.

In between the crash and the arrival of the bad news, one of the four dead students came to the party. He entered the house along with some others.

Later, when the news of the crash came, the people at the party were initially reluctant to believe it, because one of the four allegedly dead students was at the party. People had seen him come in! However, he could not be found, and no one could remember him speaking to anyone, or doing anything except walking into the house. He had arrived, probably within moments or minutes of his physical death, and then hadn't been seen again. This is a classic example of a crisis apparition.

A Worldwide Phenomenon

*You live on Earth only for a few short years which you call an incarnation,
and then you leave your body as an outworn dress and go for refreshment
to your true home in the spirit.*
Chief White Eagle

*Those places where sadness and misery abound are favored settings for
stories of ghosts and apparitions. Calcutta has countless such stories
hidden in its darkness, stories that nobody wants to admit they believe but
which nevertheless survive in the memory of generations as the only
chronicle of the past.*
Carlos Ruiz Zafón, *The Midnight Palace*

***

Read any book about ghost sightings and beliefs, and you'll find some of
the famous standards: Raynham Hall; the Borley Rectory; the
Winchester House; the Stanley Hotel; the Whaley House; Hampton
Court Palace; Edinburgh Castle.

What do these places have in common? Well, for one thing, they are
in Great Britain and the United States. Although some books may make a
point to expand the geographic location of the hauntings they discuss, for
the most part, these two countries make up the bulk of the stories
recounted. For that reason, I won't be focusing on these hauntings. I
suspect that many, like myself, are simply a bit tired of hearing about
some of the old standards (though I do intend to drive past the site of the
Borley Rectory next time we're in Britain). There have to be interesting

tales from other places, too! In fact, for this chapter, we'll look farther afield.

Hong Kong

In Chinese culture, ghosts are generally considered dangerous or unhealthy, though they are not always viewed in this way. To find out more about attitudes about ghosts, Charles Emmons and his wife at the time conducted over 3600 interviews with residents of Hong Kong via a random telephone survey during 1980-1981. Though Emmons was warned that the Chinese would not want to share their thoughts on the subject, he found that he had a very good success rate, especially as his Cantonese-speaking wife honed her interviewing skills over time.

Though the vast majority of the respondents who said they had encountered ghosts claimed to only have one experience, a few people had multiple experiences, and some even admitted that they saw ghosts quite frequently. A few quotes from these people:

"The first time I discovered that I could see ghosts was in my teens in Canton city. I saw ghosts with white tops and black slacks. They were just walking down the street...they ran right through a wall."

"I can see ghosts all the time in the house here. I don't recognize any of them. Nobody else sees them. They are all different; I never see the same one twice."

"I have seen a few figures all in white gowns at least five or six times in that house. Sometimes, I'd close the door, and they would open it for me. Sometimes, they would lock it from the inside."

Emmons also found that certain days, such as the seventh day or the one hundredth day, after death are significant.

"My sister later appeared to everybody else in the family, as a shadow, floating in the air. Because she suffered and died a violent death, she was fierce as a ghost. The seventh day after her death, she came back according to the custom..."

"*My grandpa came back on the seventh night after his death. It was very horrible. The next day, all the food, water, and candy had been eaten or moved around.*"

"*Seven days after my mother died, I was half-asleep when I saw her standing beside my bed, looking at me. She stood there for two or three minutes, and I couldn't move. All of a sudden she disappeared.*"

"*The day my grandmother's soul was supposed to return, I was not yet ten years old. That day, my family prepared the altar, including food and fruit. And then at night, my whole family got up. We went to the altar. We found that the fish there was gone, and the strangest thing is that one of the longlife buns had a handprint on it.*"

"*Exactly one hundred days after my father's death, I was working with an electric saw, which I couldn't fix. I looked up and saw my father, who was very handy with tools, smiling at me. He didn't move and made no noise. I saw him for just a couple of seconds. No one else saw him but me.*"

Being the first person to spot a ghost can be quite unlucky. One story related to Emmons was from someone who had been part of a group of people who were hanging out late at night. Suddenly, a man in a white robe appeared. "My aunt yelled, "Look who's here!" and we all saw him walk from one side of the patio to the other and go in the house. We went into the house and asked the woman who lived there who had come in...He was her deceased husband...My aunt was immediately afraid when she realized that she had been the first to see the ghost. She got very sick one week later, and died a month later."

Hungry Ghost Month

Some people may be familiar with Hungry Ghost Month. Hungry Ghost Month is the seventh lunar month, while the Ghost Festival itself is celebrated on the 15th day of that month. This ghostly time of the year has been featured in such movies as *They Wait*, where the spirits kidnap a young child, and his mother must go to the world of the ghosts to get him back.

During this month, the gates to the afterlife (this afterlife is often referred to in English as *hell* but should not be confused with the Christian hell) are opened so the spirits of the dead can return to earth. Ghosts who return to earth for Hungry Ghost Month are often those who

were not treated properly after they died. These ghosts may strike out at the living; thus, making offerings to all ghosts, not just one's own relatives, can help the living avoid being plagued by the dead.

Ghosts are offered places at the table, where a chair is left vacant for them and food is placed before them. At theatrical performances, a row of seats is left empty for the dead theater-goers. An altar with offerings will be placed on each street, and sometimes, shops will close so that the streets are available for the dead to use.

Offerings are often paper, cardboard, or papier-mache versions of the real things (money, cars, houses, clothes), which are burned so the ghosts can use the items in the afterlife. On the last night of the month, monks chant to let the ghosts know it is time to go. People may also place lanterns made of paper on the surface of a river, and the ghosts will follow the lanterns away from the land of the living.

Japan

Though Japan is home to many odd and frightening folklore creatures, many of them are things which were never human. The human ghost

A Yurei. (Wikimedia Commons)

everyone is familiar with these days is the *yurei*. They are often female, wear white kimonos, and have long black hair. If you've watched movies like *The Grudge*, you've seen a representation of a *yurei*.

Some *yurei* are looking for vengeance because something bad happened to them, and they want to make sure someone pays. Women who died while giving birth can become *yurei*. So can those who die at sea.

For over a thousand years, the Japanese have been building shrines and performing certain religious practices that are designed to appease these vengeful spirits. This custom continues today; in Tokyo, there is a shrine dedicated to the ghosts from World War II.

Interestingly, even more recently, *yurei* have been active since the 2011 earthquake (the one that damaged the Fukushima reactor). A reporter found out that people have been seeing dead loved ones, and that a local fire department has received calls requesting assistance from houses that were destroyed. I have to say, receiving phone calls from non-existent houses is definitely on the creepy side.

Yurei are immortal; you cannot get rid of them. You can only try to figure out what they want and give it to them, though even that is no guarantee. This seems to go along with the fatalism that is in Japanese ghost movies: the ghosts simply *are* and, if they're out to cause harm, there's nothing ordinary people can do about it.

Also, I need to quote this, because everyone should be exposed to this paragraph. "Both feared and revered, *yurei* are part of the deep magic; a foundational belief that humans have a god inside of them. This powerful, supernatural entity—called *reikon* or *tamashi*—is held in check only by the meat-cocoon of the body."

Meat-cocoon. Couldn't have phrased it better myself.

Thailand

The most famous ghost in Thailand is that of Mae Nak; her story is told and retold in film, opera, books, and more. She was a pregnant newlywed when her husband was conscripted into the army. Unfortunately, both she and the baby died while he was away. Yet her desire to be reunited with her husband was so strong, she became a *phi tai thang klom*—a ghost of a woman who died trying to give birth. Although in life the woman may have been sweet, the ghost can turn vengeful toward anyone who tries to break the illusion that she is alive and all is

> **Fun Fact**
> You can watch Nang Nak, a 1999 film version of the story of Mae Nak, at www.youtube.com/watch?v=-l01jEkGuhM

well. Mae Nak's husband eventually realized that his wife was a ghost, and what he thought was his son was actually the corpse of the infant. He took shelter in a monastery and had to call on assistance from the famous monk Somdet To to exorcise his wife's spirit from this world.

There is a shrine to Mae Nak in Bangkok which is very popular. People appeal to her for winning lottery numbers (she's apparently good at choosing those), to avoid the draft (which she hates since it took her husband from her), and for a safe delivery of a baby (since she understands the tragedy of childbirth gone wrong).

The most gruesome ghost (to me, anyway) is the *krasue*, which consists of a floating female head with its spine and internal organs hanging down from its neck. Sure hope I never encounter one of those! Weirdly, *krasue* float around looking for clothes left outdoors overnight. They will rub themselves on the clothes, leaving the clothing covered in bile and other secretions. In the morning, the person will discover their clothes stained with the *krasue*'s muck. At this point, the only thing you can do is toss your clothes away. Who'd want to wear them again, anyway, right?

The best way to avoid a *krasue* is to surround your house with plants that have sharp leaves. A *krasue* will avoid the sharp leaves lest their dangling organs be damaged. Apparently, they don't care for that.

The *phi tai hong* is the ghost of someone who died an unnatural death such as being murdered, committing suicide, or dying in an accident. They are bitter because they died too young and there's no way to know what they'll do to you if you should meet one. You can just be sure it's going to be bad.

Phi tai hong that die in car accidents will stay by the road and try to make others crash. To avoid this, sometimes people build shrines at the accident site. Honking one's car horn when one passes a shrine is said to show respect to any *phi tai hong* in the area. But you should still stay alert lest the ghost attempt to run you off the road and get you to join them on the other side.

Wondering if all Thai ghosts are vengeful? Well, how about a good ghost? Then look no further than the *phi nang tani*. She lives in a banana plant and will sometimes reveal herself on the night of the full moon. When she appears, she shows herself as a young woman wearing traditional Thai clothing in the color green. Her skin is slightly green but I assume that since she can only be seen in human form during the night of the full moon, the undertone of her skin may be difficult to see clearly. Her feet will hover above the ground. People who believe a particular banana plant to be inhabited by a *phi nang tani* will wrap the plant with colored cloth and will leave her offerings of incense, flowers, and sweet snacks. She will give out her bananas to those who are hungry, especially to monks.

I couldn't let this discussion go without mentioning the *phi kee*, a ghost that occupies toilets. If you have a discussion with it before you use the toilet, where you ask it to allow your excrement to pass peacefully, it can remove bad luck from you. I've never heard of a toilet ghost before, but now I guess I know who to talk to if I feel I've been afflicted with bad luck.

Under the Weather? About that...

The ghosts of things that never happened are worse than the ghosts of things that did.
L.M. Montgomery, *Emily's Quest*

Ghosts have a way of misleading you; they can make your thoughts as heavy as branches after a storm.
Rebecca Maizel, *Infinite Days*

✳
✳✳

If the ghost shows on my television are to be believed, ghosts generally aren't capable of harming people. If the ghost hunters go to a location where someone has been pushed or scratched, they're liable to start talking about demons or other entities that were never human. But that doesn't mean that ghosts are completely off the hook when harm comes to a human being. In some cultures, ghosts are thought to bring disease, or even to try to possess the living (another thing generally relegated to entities of a more demonic nature on ghost hunter shows). Researchers have laid out a few cases where ghosts make people sick, or even possess them and attempt to cause their deaths. And these are *ghosts*, the spiritual remnant of a human being (often a parent or other family member), not something else.

Northern India

A belief in ghosts is found in the religions of North India. Ruth and Stanley Freed report on one town near Delhi, in which the prevailing belief is that people can become ghosts for several reasons: they die before their time; they die due to accident, murder, suicide or of certain diseases; or, in life, they acted contrary to the customs of the village. Once dead, the weight of the soul's actions over the course of their many reincarnations determines whether or not they will continue on to another lifetime, become a ghost, or graduate from the cycle of rebirths. Adding the most recent death, if it were due to the reasons cited above, to the soul's tally may be enough to send it back to earth as a ghost.

Interestingly, the rate of belief of ghosts is dependent upon gender, with women reporting a belief in ghosts far more often than men. Children reported believing in ghosts because their mothers told them ghosts were real, and "mothers do not lie." Once grown up, many males who believed as children have stopped believing in ghosts. The Freeds consider this is due to the higher rate of education among men, the division of labor between the sexes (hence men and women rarely find themselves in the same work or social situations where they might share stories), and also that, at some point, men come to believe that ghosts are simply a reflection of a person's fear, and fear is something more common in women. One might sum this up as "belief in ghosts is girly," so to speak, though that doesn't keep some men from maintaining a belief in ghosts into adulthood.

While the belief in ghosts by women may certainly be chalked up to fear, it's more complicated than that. Women may find themselves blamed for situations, such as the death of an infant, that are out of their control. When disappointing one's husband's family can result in severe social opprobrium, up to and including physical danger, possibly even death, being able to scapegoat a ghost can be a way to protect oneself. Because sons are favored above daughters, the death of a son is often especially fraught with danger for the mother. "Shifting the blame for the death of a son to a ghost, often the ghost of an earlier wife who died, may absolve a mother from ignominy in the eyes of her in-laws."

Also, having a convenient supernatural scapegoat can help both parents relieve themselves of their own feelings of guilt, especially if the father also believes in ghosts. After all, one can fight the supernatural, but that doesn't necessarily mean one will win.

But how do ghosts harm people, and even kill infants? Villagers recognize the signs: fever, the person speaking in a strange voice, convulsions, odd bodily movements, choking, and crying (at least, in infants). Fever is the number one way to spot an illness caused by a ghost, but choking is also common, as it shows the ghost is trying to remove the victim's soul via the mouth.

Villagers have many options when they want a specialist to help them cure someone of a ghost illness or ghost possession. Pandits are members of the Brahman caste who are thought to have supernatural powers. Bhagats, siyanas, and maharajahas are exorcists from other castes. Vaids, who follow Ayurvedic Medicine, pharmacists, various village healers, and doctors trained in Western medicine are other specialists parents might visit.

Some parents practice a sort of mix-and-match approach. They may take their child to a doctor *and* a village specialist of some variety in an attempt to cover all the bases. One woman (referred to as "Mrs. Fence Sitter" by the Freeds) obtained an antibiotic from a doctor but also acquired crystals from a healer who told her to throw them in a fire and make the child breathe in the fumes. Since the child lived, and the ghost was effectively exorcised, this woman continued to practice this combination of modern medicine and folk advice with all her children.

Another woman who tried the mix-and-match approach did not have a positive outcome. When her son fell ill, she at first used traditional methods, including placing an amulet around the boy's neck, on the advice of the local specialist. However, her child did not get better. Eventually, the woman and the rest of the family took the infant to the hospital, where the doctors explained that the child was in serious condition. The child was placed on oxygen, but died shortly afterward.

While the woman blamed the *ghost* for her son's death, she avoided using *Western medicine* for a decade before deciding to go to a hospital to give birth to a later child. She continued to have faith in her local healing traditions and specialists. "The boy's death confirmed the beliefs of the village people who henceforth followed traditional practices."

Besides giving people substances to burn so they can breathe in the fumes, and amulets to wear on a necklace, specialists might also recite mantras to help exorcise the ghosts causing the fever. If the exorcist considers it necessary, the victim may be slapped or have their hair pulled so that the ghost, shocked or insulted by the abuse, will simply go away. But even if the ghost agrees to leave a body and stop the fever, the family may not blame the exorcist if the person gets worse, or even dies, because ghosts are considered to be liars. The belief is that "ghosts do not keep their promises." The exorcist is absolved of responsibility for negative outcomes.

In short, ghosts are a convenient way to eliminate guilt, avoid danger to one's own safety from grieving in-laws, and to explain how someone who was being treated by local healers and/or Western medicine could continue to get worse, or even die, despite the family following every instruction they were given. Ghosts are angry and vengeful, and are clearly capricious beings that will say and do whatever they wish. Humans have some ability to thwart them, but even the best healer's medicine or advice is not a guarantee that ghosts can be defeated by the living.

North America

One of the dangers of ghosts is that they may not wish to walk off into the sunset by themselves. They may think that a companion would be just the thing to make the road to the afterlife a bit less...dark? Scary? Lonely? In any case, to that end, some cultures have prayers said over the dead that implore them to leave the living alone. Robert Putsch quotes several Native American funeral prayers that were recorded by other researchers.

> Don't seek us and we won't seek you;
> Don't yearn for your relatives.
> Don't call for us.

> Tell them not to trouble us, or not to come here
> And take anyone else away.

Go, go straight ahead
Do not take anyone with you.
Do not look back.

It is a well-known phenomenon that some people, when bereaved, will feel a self-destructive impulse which can lead to suicidal ideation or a lack of self-care. In one study, a full 65% of the participants agreed that they had experienced thoughts of wanting to die after the loss of a loved one. More than half also indicated they had actively engaged in self-destructive behavior. While only 9% attempted suicide, 29% reported "indirect suicidal behavior." Indirect suicidal behavior can include such behaviors as driving recklessly, smoking, refusal to take one's medications, not looking before crossing the street, or other behaviors which do not directly constitute suicide, but which may make injury, disease, or death more likely.

There's actually a psychological term for this: *the call of the void.* Have you ever stood on the edge of a cliff and contemplated what it would be like to jump off? You didn't follow through, but that didn't keep your brain from wondering about it. Theories as to why this happens include a survival instinct (you recognize the danger, so you then avoid it) or a kind of affirmation of life (as if your brain says to itself: *see? it would be dangerous to jump, so you aren't going to do it*). Some of those who are grieving may experience this call of the void more than usual.

In one example, the researchers related the case of a Navajo woman who was in some distress two months after giving birth. She had experienced crying spells, insomnia, a lack of interest in her regular activities, and frequent spats with her spouse. On two occasions, she had been so angry, she had gotten in her car and driven 80-90 mph. After the second of these incidents, she had been frightened enough to go to the emergency room to ask doctors why she was "losing her mind" and "not caring about anything."

On top of this, the woman reported dreams of her late father, which "made me feel like I'm going to do something crazy." The woman's father had died over six years earlier; her mother had advised she have a ceremony to get rid of the influence of her father's ghost. However, the woman's husband, a Catholic, didn't believe in native ceremonies.

The researchers mention that part of the issue is that Navajo culture doesn't really consider any death "natural" outside of death from extreme old age. People don't just die of disease or accident; they are *killed*. Therefore, someone engaging in self-destructive behavior is not harming themselves, but is the victim of an external force. In this case, the woman and her family agreed it was her father "driving you to it." The researchers noted that the woman's mother had not, in diagnosing the issue, focused on the woman's isolation in a small mountain town, the new baby/post-partum issues, or the fact that the woman and her husband were now fighting a lot. Instead, the explanation jumped right to a dead relative, in this case, the mother's own late husband, as the cause of the dysfunction in the family.

Eventually, a treatment plan was developed which included a trip to the Navajo reservation for ceremonies designed to get rid of the malignant influence of the ghost. This resulted in an improvement in the woman's condition.

In a second example, a Salish woman was afflicted with crippling arthritis. Like the Navajo woman, she had also dreamed of her late father. In this dream, he had given her a spirit song. Unfortunately, the owner of a spirit song can become possessed by the ghost that has targeted them. Those who are possessed may, again, experience thoughts of self-harm. Consequently, the woman had told her mother that if she couldn't get rid of the controlling spirit by spring, she would end up going with her father. She visited several different specialists but no one could help her. One told her that the spirit was too powerful for him to deal with. Another noted that the woman had lost her soul in the cemetery at the time of her father's funeral. This correlated with the woman's assertions that an external force had caused her to want to jump in the grave with her father and that this self-destructive feeling had stayed with her.

The woman believed that, at the second anniversary of her father's death, his soul would move on "and become less of a threat to the living." She greatly feared dying before then, but she and her mother were alive to participate in a memorial service for her father. After this, while the woman's arthritis continued to be a problem, her psychological health improved significantly.

A Laotian in the U.S.

A young Hmong woman living in the U.S. began experiencing terrible headaches which sometimes lasted as long as 48 hours. Sometimes, the headaches were accompanied by nausea and vomiting. She had visited emergency rooms on multiple occasions and had been prescribed various medications, but nothing had helped. Her husband reported to the researchers that his wife sometimes woke up from dreams of her late parents to announce that she was going to die. He said, "She thinks he's [her father] going to take her with him."

Though the woman felt she needed to find a way to remove her father's spirit's influence over her, she insisted that, "as a Catholic," she didn't know what to do. Both she and her husband were very recent converts to Catholicism; the researchers felt that the couple's insistence that they "didn't know" what to do because of their religion was an attempt by them to fit in with their new community. Though both were originally from Laos, they had met in a refugee camp in Thailand before being brought to the United States. They did have family members living nearby, but ultimately, they were living in a country where they did not speak the language or have much familiarity with the culture. Their feelings of isolation and alienation were very strong. Clinging to their new religion was viewed by the researchers "as attempts to avoid being labeled as different."

The young couple consulted an uncle to learn more about the situation. The uncle, as well as other elders of the family, were then able to offer the diagnosis of a ghost illness and an explanation for why it was happening: the couple had not asked their parents' permission to be married.

Once the problem had been diagnosed, family members almost immediately rallied to prepare a meal, gifts, and prayers for the deceased parents of both husband and wife. After offering these things to the parents, the couple's fortunes changed for the better. The researchers discovered, in a follow-up visit, that the young woman "had become cheerful, animated, and involved. She remained headache-free for a six-month period after the meal."

Ghosts Make You Sick

These examples indicate that, in some cultures, the ghosts of deceased relatives, especially parents, are capable of possessing their descendants in an attempt to take them to the afterlife with them. While, to Western eyes, it may look like the bereaved and their communities are navigating grief in ways that bring the community together, that is not necessarily the understanding those involved have of what's happening to them. Once participating in some kind of ritual meal or other ceremony, ghosts are exorcised and the living can stop being quite so oppressed by their grief, their fear of imminent death, and their tendency to engage in self-destructive behavior. Physical ailments, such as arthritis, may remain, or they may, like the Hmong woman's headaches, be ameliorated, but ultimately, the physical issues are secondary to becoming free of the malignant influence of the recently dead.

The other thing these ghost illnesses may assist with is giving people a way to discuss their issues that their cultures otherwise have no way to discuss. The Hmong refugees experienced extreme trauma in losing their homes and their parents, and being forced to live in refugee camps. Then they came to a strange country where they could not communicate and did not have good coping skills for adapting to an alien culture. Add to this, the Hmong do not have terms for conditions such as anxiety and depression in their culture. Without being able to discuss their psychological issues within their own culture, the couple still had the option of experiencing a ghost illness that would cause their families to rally around them and to support them. The Salish woman and the Navajo woman experienced similar social support via their need to be relieved of the influence of ghosts on their own minds and bodies.

Does that mean these ghosts are only in people's heads, and have no reality outside being a culturally appropriate way to cope with depression, anxiety, and grief? Ultimately, that is a question I can hardly answer. I could certainly discuss being anxious or depressed, but could I go to a friend or doctor and claim I'm being made ill by a ghost? That sounds like a good way to be told I'm inventing a ghost as a scapegoat because I don't want to confront my *real* mental health issues.

Your culture may vary.

Across the Veil

After your death, you will be what you were before your birth.
Arthur Schopenhauer

I believe as humans we are blessed with a soul. Our life force is so strong, complex and well-developed that I find it hard to believe that it only popped into existence when we were born and on death it just ends.
Richard Stuttle, author

Life is millions of billions of trillions of transformations of something into some things, or another thing; never into nothing.
Mokokoma Mokhonoana, author

**

If humans survive death in some form, if we are still ourselves even after our bodies cease to function, then where is it we go, or what is it our souls experience? Cultures have had various answers to this question, from dark realms of dread and terror (or even eternal boredom), to blissful fields of flowers, to staid realms of prayer, to wretched places of torture and punishment. Sometimes, one's placement after death depends upon one's actions in life. Other times, there is simply a common fate for all humans, whatever one's actions or beliefs during one's lifetime.

In the *Epic of Gilgamesh*, Enkidu has a dream of the afterlife that indicates it is, basically, the grave. After death, one exists where there is nothing but darkness and dirt. The afterlife, in Sumerian, even glosses into English as *House of Dust*.

This rather unsophisticated view of the afterlife became more elaborate in later civilizations. The Greeks wrote about lands like Hades and Elysium and Tartarus. The afterlife could now be pleasant or unpleasant, depending on what sort of life one lived on earth. Those who were especially heroic might be raised to the heavens to become a constellation and shine down on the earth forever.

The Mayan afterlife, as described in the *Popol Vuh*, was a dark place full of monsters and other terrors. Today, the star Polaris is the northern pole star, but when the Mayans were recording the sky, they realized all the stars rotated around a dark void. This void, the Black Transformer, was the way to get to the Underworld, Xibalba. The underworld had many levels and was ruled by deities with names like Seven Death, Pus Master, Blood Gatherer, and Jaundice Master. If these terrible deities came to Earth, which they did on occasion, they spread disease and other terrible things. Getting through this underworld meant facing fearsome rivers, high mountains, attack by obsidian blades, even the loss of one's heart.

> Lots of afterlives seem to have terrible fates in store for many humans. But not every culture has believed in the meting out of punishments, or even the bestowing of pleasures. The characters in *Beowulf* have no belief in such a thing. The epic advises anyone who wants immortality to be remembered for their heroic deeds. The poet says, *Let him who can win glory before death. Once a warrior is gone, this is his best and only bulwark.*

The dead were buried with grave goods which could help them survive these afterworld perils (including dogs, which might guide them along the way). Depictions on pottery show the dead suffering while their flesh drops off their bones or their eyes fall out of their sockets. This afterlife is truly a frightful place. The Hero Twins of the *Popol Vuh* survive by tricking the death gods into killing them in such a way that they will come back to life. However, since no one can undergo death and return without being transformed, the twins return as the sun and the moon. But the tale of their survival, their cleverness in tricking the death gods, their resurrection, and their return to Earth must have given people hope, despite the horrific descriptions of the afterlife that had come before.

Of course, the afterlife that most people reading this are probably familiar with, in some capacity, is the idea of heaven and hell. The more mainstream ideas associated with these concepts have developed over the centuries. Neither is found to any great extent in the Old Testament, but during the first few centuries of Christianity, the ideas began to coalesce into more definitive doctrine.

Not that there's agreement among all Christians on doctrine. Though 1 Thessalonians 4:16-17 says, *"For the Lord himself will come down from heaven, with a loud command, with the voice of the archangel and with the trumpet call of God, and the dead in Christ will rise first. After that, we who are still alive and are left will be caught up together with them in the clouds to meet the Lord in the air. And so we will be with the Lord forever,"* most Christians I've met insist that the moment someone takes their last breath on Earth, they take their first breath in heaven. But, ultimately, there is still a belief in a heaven, even if there is disagreement on when souls might actually end up there.

So one might concede that ghosts would be easy to incorporate into Christianity if one assumes souls do not go directly to heaven upon the death of the body. That would leave a lot of souls with nothing to do until the day *the Lord himself will come down from heaven.*

Hell, of course, has been of far more interest to most since the invention of the concept. Is it because contemplating the dire punishments caused by disobeying God is an appealing hobby? I suspect it's at least partly due to the fact that people in past centuries did find enjoyment in watching others suffer, at least more than we do now (or most of us). We can see this when looking at medieval artwork and note that the blessed are worshiping God in heaven while simultaneously glancing down to watch the torments of hell (a habit known as *abominable folly*). Observing the suffering in hell, *and enjoying it* (even though one must assume some of those experiencing the torments of hell include the blessed souls' parents, siblings, spouses, and children), makes the blessed seem like horrid psychopaths. Who could watch their loved ones suffer and think it's entertaining? *Who?*

This very short survey of a few afterlives give us just a hint of the vast array of beliefs from human civilizations as to what happens to us when we die. The important thing is that, as far as I know, all human societies have had some kind of belief about an afterlife. And, of course, there'd be no need to have an afterlife if one didn't believe that some portion of the human being survived death.

Can Souls Be Captured or Measured?

This question has been on the minds of at least some people for centuries. Perhaps due to the King James Version of the Bible describing Jesus' last breath as "giving up the ghost," the idea among many people was that the soul escaped the body with the last breath. Interestingly, some sources claim that Thomas Edison believed this and thus asked for his friend Henry Ford's last breath with the idea that he would then be in possession of Ford's soul. Whether or not this is actually true is not known with certainty but the story does indicate that the idea of linking the last breath with the soul is still around.

Others weren't worried with capturing the last breath or the soul as much as they wanted to figure out when souls (assuming they existed) left the body.

Readers of this book may already be familiar with Duncan MacDougall, who decided to try to measure the weight of the soul. To accomplish this, MacDougall weighed people before and after death. One of his subjects lost 21 grams at that time. To MacDougall, this was proof of a soul and also proof that we could use scientific methods to measure it.

Duncan MacDougall
(Wikimedia Commons)

MacDougall's experiments have been dismissed for several reasons, including his small sample size (six), and the fact that only one of those people lost the 21 grams at death.

Yet the idea that, no only do souls exist, but that they exist in a way that can be sensed by our tools, even simple ones like a scale, was a natural outgrowth of the scientific revolution. Want to discover something? Find a tool that might help you measure it. Point a camera at it. Wrap your calipers around it. Modern ghost hunters have taken this idea to heart in a big way: they've gone from scales to cameras, EMF meters, ghost boxes, thermal cameras, digital thermometers, vocal recorders, and just about every other gadget you can think of, but the motivation is the same.

Still, no one has ever actually been able to show that humans have some kind of substance inside them that persists after death. That doesn't keep most humans from believing in the existence of a soul. And *that* doesn't keep people from telling interesting stories of what happened to them after they died and were resuscitated.

Heading Across the Divide to...What?

I wasn't anything. It was just like this raw form of consciousness where I was just existing very happily and pleasantly. I just started fading into the fabric of the universe. It was so warm and peaceful and pleasant.

Adam, Near-Death Experiencer

He said that I've gone as far as I can, and if I go any further, I won't be able to turn back. But I felt I didn't want to turn back, because it was so beautiful. It was just incredible, because, for the first time, all the pain had gone. All the discomfort had gone. All the fear was gone. I just felt so incredible. And I felt as though I was enveloped in this feeling of just love. Unconditional love.

Anita, Near-Death Experiencer

The next thing I knew was there was something behind me, and I was afraid. I felt this awful presence—and I knew that it was after me. This...this thing, this awful, terrifying thing—I could feel it on me. I was being chased by something that was the personification of evil. And it wanted me. It wanted to destroy me. I was terrified and I was crying, and I remember thinking: oh God, help me, help me, God.

Dee, Near-Death Experiencer

The concept of the near-death experience (NDE) has become so well-known that few, if any, would be surprised to hear that someone who had been clinically dead for a few minutes and then revived, had some kind of story to tell about what happened to them while they were technically dead. I have quoted the people above to show that not everyone has exactly the same experience; in fact, some percentage of people have frightening experiences where they are tormented, chased, or otherwise made to endure terrible fear and pain before being sent back to their body. These experiences do not make up the bulk of NDEs but some suspect negative experiences are underreported due to the fact that "everyone knows" that NDEs are peaceful experiences where one sees a white light, is enveloped in a feeling of love, meets loved ones who have died previously, and are told at some point to turn back, which they don't want to do. Some researchers wonder if people keep negative experiences to themselves because they know the standard script and are either embarrassed by or worried that their experience will not be taken seriously since it's not what other people expect to hear.

Also, having a frightening experience and being someplace you and your friends and family are liable to label hell, could be seen as embarrassing or shameful. After all, one wouldn't be sent to hell if one were a good person. Admitting one has been to a place of punishment may be seen as also admitting that one has not led a good life. Does anyone who has survived death really want to be labeled a bad person at the same time?

Arvin Gibson found that those who had experienced the frightening NDEs had already known they were leading less-than-savory lives. They were hooked on drugs, trapped in abusive relationships, had attempted suicide, or were even murderers.

Oddly, the murderer was only nine years old when he had his NDE. Gibson said, "I could not understand why a little child would be subjected to this kind of evil influence—until he told me the rest of the story. It turned out that when he was about six years old, he and another small friend were fighting. Mike picked up a piece of lumber that had a large nail sticking out of its end, swung it at his friend, hit him in the head, and killed him. He tried unsuccessfully to hide the body. The police later dismissed the case as an accident."

The police may have considered this an accident, but clearly Mike knew what he had done was wrong, otherwise, he would not have tried to hide the body. So that's how a six year old murderer turned into a nine year old with a negative near-death experience.

Most people who tell of their NDE experience something much more peaceful and pleasant. I suspect most people these days know the basics: a feeling of leaving the body, a cessation of pain, a white light, a feeling of life review, the presence of loved ones, and a feeling of great peace. Sometimes, instead of, or in addition to, loved ones, the experiencer feels they are encountering beings of light. At some point, the experiencer is either given the option of returning, and decides to, or is told to go back because it's not their time. They then suddenly discover they are back inside their body.

Sometimes, while separating from the body, people report hearing conversations that their doctors or other people around them are having. They believe they can observe what is going on with great clarity before being drawn into the light and moving away from this plane of existence.

While most doctors believe the NDE is a product of the shutting down of the brain, the experiences do show how people might have developed more detailed stories about what happened after death. All it would take is for a few people return from a death-like state to tell their story about lights and encountering dead relatives before significant beliefs in ghosts and their place in both this world and the next would be developed.

I have no doubt that an NDE is a very real experience, but is it merely the experience a dying brain gives itself to try to make sense of portions of it shutting down, or could it be a genuine experience of a soul leaving the body and heading out into the universe?

And could it mean that those who refuse to go to the light could actually stay on this plane and wander the world plaguing the living? That would explain a lot.

Ghost in the Brain

*It's possible that the reason I've never experienced a ghostly presence is
that my temporal lobes aren't wired for it. It could well be that the main
difference between skeptics and believers is the neural structure they were
born with. But the question remains: are these people whose EMF-
influenced brain alert them to "presences" picking up something real that
the rest of us can't pick up, or are they hallucinating? Here again, we must
end with the Big Shrug, a statue of which is being erected on the lawn
outside my office.*
Mary Roach, *Spook: Science Tackles the Afterlife*

*Throughout history a handful of researchers have dedicated their lives to
discovering what supposedly paranormal phenomena tell us about our
behavior, beliefs and brain. Daring to take a walk on the weird side, these
pioneering mavericks have carried out some of the strangest research ever
conducted...Just as the mysterious Wizard of Oz turned out to be a man
behind a curtain pushing buttons and pulling levers, so their work has
yielded surprising and important insights into the psychology of everyday
life and the human psyche.*
Richard Wiseman, *Paranormality: Why we see what isn't there*

✳
✳✳

Those *H. heidelbergensis* individuals who dropped their dead down the
chute may have been among the first proto-humans to worry about death.
We see the same concern in the Neanderthals of Shanidar. How would
death come? When would it come? But, especially, what would happen

after that? The body died, but surely the mind must survive somehow. Right?

That question has haunted people ever since. Otherwise, self-proclaimed psychics like James van Praagh, Sylvia Brown, and John Edward wouldn't have book deals, television shows, and thousands of adoring fans. Interestingly, while James van Praagh currently has 542K followers on Facebook, John Edward has only 122K (far fewer than I would have thought). Edward has 118K followers on Twitter, very similar to his Facebook presence, while van Praagh has a mere 58K, only 1/10th the number of his Facebook followers. Make of that what you will.

Sylvia Brown has passed on, which may be why she does not seem to have a Facebook or Twitter presence (though she's got fan groups on Facebook). Perhaps her estate eschews social media.

It's not hard to find psychics, astrologers, and other spiritualists who are more than willing to charge you to develop your talents or to simply answer your questions about love, money, or your late loved ones.

The obsession with wanting to know about life after death can lead people to accept almost anything, especially when they're desperate. If someone feels guilty or otherwise can't cope with the death of their loved one, they can be extremely vulnerable when presented with someone who claims to be able to let the communicate with that loved one.

But why are people so easy to fool? Is it just grief that makes people vulnerable? Or are there other factors at play?

Ease of Fooling the Senses
One of the things that people like to depend on is our brains. Sadly, our brains are not as dependable as we like to believe they are. Brains are notoriously easy to fool, a trait which magicians and other tricksters take advantage of every day.

Consider those little drawings you may see in a newspaper (or online in this largely post-print newspaper days) that ask you to "spot five differences between these two drawings." Can you do it? I usually can, but it can take a while. The trouble is that the brain, once it sees something, sticks with what it has already seen without updating itself when new information comes in, especially if that new information is rather subtle. Also, our brains do not register everything we see: to do so would simply take too much processing power and too much memory. Instead, the brain focused on a few things, and lets the rest go. No matter how many times book or movie characters talk about needing to "access memories" that they are sure they have but just can't get to for some reason, those memories don't exist. The brain is not recording like a mechanical device. It is making quite a few decisions about what to pay attention to and what to dismiss and it doesn't bother to let you know.

Modern Spiritualism & Psychics

Here's another spot where I'm going to assume most people reading this book have a great deal of knowledge and will only give the basics. Modern spiritualism began in 1848 New York when the Fox sisters began communication with spirits via raps. Could the spirit count to five? Five raps would be heard. The sisters had the good fortune of living in an area where people were following all the latest religious and spiritual fads (like Millerism), and they soon found themselves at the center of a growing movement. They went on tour and, when the huge numbers of casualties of the American Civil War increased the number of dead loved ones that people wished to communicate with, they found themselves with more followers than ever.

Even though the sisters later came clean and explained how they had performed all the tricks themselves and proclaimed they had never actually been in contact with the dead (though one of them, Maggie, later recanted her recanting), it was far too late for the truth to make a difference. People were too invested in the idea that their loved ones, even though dead, were not wholly cut off from the world and communication with the living. Grief could, perhaps, be ameliorated, at least in part, by the comfort that could be gained by sending and receiving messages to the other side.

Spiritualism came up with a variety of methods to help this communication, including lampadomancy (flame reading); demography (skin writing); psychometry (getting impressions off physical items); scrying (the old crystal ball trick); contacting spirit guides; spirit photography; spirit rapping (the Fox sisters' specialty); table tilting; clairliene (spiritual smells); clairaudience (spiritual sounds, including voices); automatic writing; mirror writing (writing backward); ouija boards or other spirit boards using a piece of smooth wood and a pointer (planchette) of some kind; and tasseography (reading tea leaves); among other techniques. Though some of these have gone by the wayside (does anyone really do table tipping anymore except ten-year-olds at slumber parties?), many of these techniques are still in use today.

Perhaps the most popular techniques today are cold reading and hot reading. Cold reading involves guessing ("I'm getting an M. Does that mean anything to you?") and following up on hits while rapidly moving past the misses. If a psychic feels the miss might be remembered ("I've never known anyone named Mabel") the psychic might play it off as having significance in the future ("Well, I'm getting that Mabel is important. Maybe that's someone you'll meet later."). That way, the psychic can never truly be wrong, even if their guesses don't pan out or their subject insists that they can't make the vague hints and clues fit their own life and situation,

Hot reading involves doing one's research ahead of time, assuming one has the option, and then being able to miraculously bring up information about a person that they don't think you're going to know. To try to avoid accusations of hot reading, psychics on television shows often mention to the viewer that they are going into a situation without being told anything about it. Some psychics have become infamous for having people fill out informational cards ahead of time so that later they can be fed information through an earpiece. Others have their staff mingle in a crowd before the show to overhear things or engage people in conversation. This information can then be revealed by the psychic on stage, as if they had suddenly gotten it from spirit guides or other supernatural forces.

With television shows, editing can help a lot. A five hour session in a studio can be edited down to the best twenty minutes. Curiously, even editing a session down to only show a small fraction of the psychic's guesses can't make a psychic's questions terribly relevant to the people in the audience. What was edited out must have been even less impressive! But people self-edit as well. If one is primed to believe, then anything that is even remotely close to what one wants to hear will be remembered, whereas every time the psychic guessed wide will not be remembered. The psychic can count on the subject to cooperate if they're already desperate for some proof of an afterlife or that their loved one is still watching out for them.

Of course, while the fact that, so far, no one has managed to be a convincing psychic doesn't mean someone, sometime, might actually be able to communicate with the dead. Assuming the dead actually exist and can be reached, and especially that they would even care about communicating with the living. I'm reminded of the story of the dragonfly nymphs who kept seeing their fellows leaving the world by climbing up the grass stems that were rooted in the stream where they lived. Their companions would then breach the barrier of the water's surface. These friends never came back down the stems to tell them what was beyond. The frightened nymphs promised that whoever was next would make sure to come back to tell the others what happened.

Eventually, one of them felt a pull toward the surface of the water. He climbed the grass stem beyond the surface, broke out of his old exoskeleton, and spread his new dragonfly wings. He looked down at the surface of the water and realized he could not go back under the water without drowning, and also that his friends would not recognize him in his new form in any case. He regretfully realized enlightening his friends was impossible. He had no choice but to break his word and leave them in ignorance. So he turned away from the water and took to his new home in the skies.

If we assume that souls exist and continue on after death, how can we also assume that these souls would even find it *possible* to continue to communicate with this plane? Why would they be reduced to rapping or leaving messages in tea leaves, or giving terribly vague hints to psychics like there *might* be some reason that *possibly* the letter M has *some* significance? Why should they not be like that dragonfly, transformed into something so different that, even if they wished to offer a message of love and reassurance, they have no choice but to turn away and stretch their metaphysical wings and take to those new skies?

Clearly, I'm not a believer in psychic phenomena, and though it would be intriguing to find someone who actually seems to have legitimate psychic abilities, as far as I can see, we are still waiting for a person like that. While I'd like to be open to the idea of ghosts or some kind of continuation of life (or at least consciousness) after death, and I'm open to people's stories, that willingness doesn't extend to psychics. Some may truly believe they are helping people contact their loved ones, but otherwise, it would seem that many are merely grifters. Accepting money and adulation from people you *know* you are conning is not a noble endeavor, and I'll leave it at that.

Just ~~Do~~ Suggest It

One of the easiest ways to deceive someone, or to make their brain deceive them, is to suggest something to them. People are terribly suggestible. This is the main reason my husband and I once left a ghost tour. The tour guide, instead of giving us the spooky history of the town, insisted on having us stand in the dark while she suggested things to us. "People who stand here often feel themselves being touched!" Or "People feel cold spots in this area!" After fifteen minutes of this near-continual prompting, sure enough, some people began to feel something brushing by them. Some felt cold. At this point, we'd had enough, and left. I had hoped to hear the ghost stories that were passed around town. What were the hauntings like? Who had experienced them? Where did these things happen? How long had it been since someone had reported an experience? Instead, what we got was an experiment showing what happens when you tell people what they're likely to feel.

Richard Wiseman describes an experiment he conducted with twenty people, where he uncorked a small bottle and announced that, soon, people would be able to smell peppermint. He asked people to raise their hands as they smelled it. Pretty soon, several people had their hands raised. Eventually, about half the people had smelled the peppermint. Of course, the bottle was empty. The smell had only been suggested.

The 1970s, the BBC performed a similar experiment during a live broadcast. The host displayed a funky contraption and announced that, when he turned it on, there would be a sound. The technological marvel of the device was that this sound was a specific vibration that would transfer a smell into people's homes, and that this smell would be something pleasant and related to the country. He turned the device on for ten seconds. After that, he encouraged viewers to call in to the BBC to report what they had smelled. A few hundred people called in to say they had smelled hay or flowers. "Many respondents described how the tone had brought about more dramatic symptoms, including hay fever attacks, sudden bouts of sneezing, and dizziness."

In 1970, in London, an associate editor of a magazine dealing with the paranormal, and a friend, brainstormed a similar experiment. They dreamed up a story about a vicar, who, in the 1800s, had beaten people to death with his cane and thrown their bodies into the Thames. This had all happened near Ratcliffe Wharf. They then went to a pub in that area, told the story, and left.

Three years later (showing that they had far more patience than me), the staff of a BBC documentary went to the Ratcliffe Wharf area and began asking if anyone knew about the Phantom Vicar of Ratcliffe Wharf. To their surprise, it didn't take long for stories of the phantom vicar to come tumbling out of the locals. One woman had seen him and even felt him watching her undress at night. A man said his children had come to stay with him for a time and had had frightening encounters with the vicar while they were there. One child in the area was so terrified of the horrific haunt that he couldn't sleep. He screamed and pointed to a corner, where his mother spotted the ghostly vicar staring at the two of them. Workmen said they had seen the vicar walking down the local streets, though he would quickly disappear once spotted.

Clearly, the experiment had been wildly successful. After a single telling of the original tale, it had only taken some repetitions of the story by the locals to have people spotting the vicar on the streets and in their homes. Even when they couldn't see him, they could sense him watching. The most tragic case, where the screaming child couldn't sleep for the horrible spectral creature staring at him while he was in bed, shows how easy people are to frighten, and how suggestible people truly can be. I wonder, if the two people who had devised the original tale had known that, within three years, the vicar would be real enough to have children screaming and sleepless, would they have done it?

Faces & Free Will Everywhere
One particular form of suggestion is one I've actually had my students have some fun with. Humans, we have discovered, come hard-wired to recognize faces. This is a useful survival skill in a species whose newborns are helpless. As an infant human, your most important job is to

figure out who is caring for you and engage them in ways that make it more likely they will continue to care for you. Because you are so helpless, and it takes such work to get you raised to the point where you can start doing things for yourself, it would make sense for your caretaker to chuck you into a bush and abandon you the first moment they figure that they'll never make back their return on investment of time, energy, and emotion. To keep this from happening, you have to be irresistibly cute and appealing. Part of the way you achieve this is by smiling at your caregiver and by making eye contact. But your eyesight isn't that good yet; you need to figure out what you should be smiling at, and you need to do it almost right away.

Pareidolia in the workplace restroom. This dispenser appears shocked that its picture is being taken.

Humans are champs at this; in fact, the ability to spot faces (even when there aren't any) is called *pareidolia*. The upshot is that you are primed by millions of years of evolution to find faces. Can't clearly see what's in your closet or just outside your window? Thought you saw something out of the corner of your eye? Maybe it's a face! A face of what, though, now *that* is the question.

Look around the room you're in right now. I bet you can spot a face somewhere. For my students, I send them out on campus with the assignment to take a picture of a face that isn't really a face. They can all do it. It isn't even hard, but it is fun.

Another thing humans are champs at is assigning agency. We are so good at it, we can do it with inanimate objects without hardly any effort at all. Wiseman discusses an experiment from the 1940s where people were shown a short cartoon of a small triangle, a large triangle, and a circle. People were then asked what was going on.

Most people instantly created elaborate stories to explain the cartoon, saying, for example, that perhaps the circle was in love with the little triangle, and the big triangle was attempting to steal away the circle, but that the little triangle fought back, and the small triangle and circle eventually lived happily ever after. In short, people saw agency where none existed....many people are very reluctant to think that certain events are meaningless, and are all too eager to assume that they are the work of invisible entitites. As such, ghosts are an essential part of our everyday lives.

That Mary Roach Quote

What do your temporal lobes have to do with sensing the paranormal? A lot, if Michael Persinger is right. Persinger and fellow researcher Stanley Koren devised a helmet that they used to generate magnetic fields around a human head. The fields were very weak, somewhere between a cell phone and a refrigerator magnet in strength. Persinger found that many people reported having mystical experiences while wearing the helmet.

Persinger's experiments have not been replicated, though Persinger argued that those whose experiments did not replicate his findings were flawed. However, some researchers found that they *could* replicate his results, except that they used fake helmets or even real ones that were not switched on. The fact that people still felt some kind of mystical experience, even when there was no magnetic field, indicated to those researchers that people *expected* to feel something, and so they did.

We're right back at the power of suggestion.

However, temporal lobes may still be the origin of many alleged paranormal experiences. Those who have temporal lobe epilepsy have reported strange experiences which have a lot in common with sensing ghosts or other unseen presences or in having strong religious experiences. Some researchers decided to "mess with" (yes, that's the technical term used in the article) the part of the brain that helps you orient yourself in space, and discovered that, while being messed with, individuals who should have felt that the signals they were receiving were coming from their own brains, instead felt as if the input were coming from outside. Olaf Blanke, the head researcher on the project said, "When the system malfunctions...it can sometimes create a second representation of one's own body, which is no longer perceived as 'me' but as someone else, a 'presence.'"

This sounds similar to the Third Man Factor, which is a phenomenon wherein a person who is in deadly danger hears a voice—an *external* voice, mind you—telling them what to do. This voice can seem male or female, it can come from over the right or left shoulder, or it can have no discernable gender or location. It will tell someone what to do until they have escaped the situation. Mountain climbers, scuba divers, and hikers have all reported this phenomenon.

In 1916, Edward Shackleton and two others hiked across South Georgia to reach a whaling station where they hoped to find rescue for the companions they had left behind (for more details on Shackleton's adventures, simply Google him). The hike was extremely dangerous and consisted of thirty-two miles through mountainous terrain. At one point, the three men could not see down a steep mountainside due to fog, and realizing they had no other choice, they simply slid down to the bottom in the fog, knowing that at any moment, they might be pitched into a crevasse or off a cliff. The men survived, but it took years for them to admit to each other that, while on this journey, they had sensed a presence with them. They could not see this fourth person, but they had all independently come to the conclusion that a fourth person was with them nonetheless.

T.S. Eliot took this story and cut the number of humans to two, with the ghostly presence making up the Third Man ("Who is the third who walks always beside you?"). Hence the name of the phenomena.

Tempted to ignore the Third Man? Probably not a good idea if you want to live. One mountain climber kept a journal of his climb and wrote that he had heard a voice telling him to get off the mountain. He mentioned in the journal that he was going to ignore the advice as the summit was so close. He died. But due to his journal entries, we know he was warned.

Ultimately, the Third Man seems to show up when a person is *in extremis*, often when they're alone, and always when they are in significant danger. Researchers posit that the brain can literally "throw its voice," so that the advice one hears seems to be coming from an external source, because humans respond well to authority most of the time. Especially when that authority is calmly telling you how to get out of a dangerous situation.

It's possible that people with the "right temporal lobes" (unlike Mary Roach) are more prone to having experiences like Blanke's team were studying, or are more likely to encounter the Third Man when they are in trouble. In past centuries (and even today), it is possible these experiences were and are chalked up to gods or ghosts.

Many people may still consider these experiences to be paranormal. And, who knows, perhaps, if you have the "right" lobes, you are easier for the paranormal to contact, and it's not your brain throwing its voice after all, but a guardian spirit that really wants to see you delivered safe and sound out of a bad situation.

So It's Just Your Brain, Then

I don't suppose you have to believe in ghosts to know that we are all haunted, all of us, by things we can see and feel and guess at, and many more things that we can't.
Beth Gutcheon, *More Than You Know*

To be haunted is to glimpse a truth that might best be hidden.
James Herbert, *Haunted*

Any phenomena that violates well-established physics...requires extraordinary evidence to support its claims.
William J. Hall and Jimmy Petonito, *Phantom Messages*

Those who believe without reason cannot be convinced by reason.
James Randi, Magician & Skeptic

✲✲

Are ghosts just all in our heads? And no, I don't mean, *sure, because all our experiences are just in our heads*, smart-ass. What we really want to know is, do ghosts have some kind of external reality?

Is there even a really good way to find out?

The Ghost Hunter

Your primary role as a ghost hunter is to prove the existence of the paranormal...

This quote from an online site that promises to certify you as a paranormal investigator bothers me quite a bit. Who thinks the primary role of an investigator is to come to the conclusion they're biased toward in the first place? That's really not much of an investigation. Would you ask someone who clearly likes *Star Wars* more than *Star Trek* to "investigate" which is objectively better, and expect them to do anything other than find some website that already confirms their thoughts on the subject? I don't think so.

As far as I can see, the primary role of a ghost hunter is to *find out what's really going on*, even if it's just a squirrel in the attic. I'm reminded of a video where Carrie Poppy discusses her own experience with a ghost. I'll quote her at length here because it's a fascinating story:

> One night, I was sitting there and I got this really spooky feeling, kind of the feeling like you're being watched. I thought, OK, it's just my imagination. But the feeling kept getting worse, and I started to feel this pressure in my chest, sort of like the feeling when you get bad news. Over the course of that week, this feeling got worse. I called my best friend and said, "I know this is going to sound crazy, but I think there's a ghost in my house, and I need to get rid of it." She said, "I don't think you're crazy. I think you just need to do a cleansing ritual." So I said, "OK." Nothing got better. So every day I'd come home and you guys, this feeling got so bad that—I mean, I'm laughing at it now—but I would sit there in bed and cry every night. I even went to a psychiatrist and tried to get her to prescribe me medicine, and she wouldn't. So finally I got on the internet, and I Googled "hauntings." I came upon this forum of ghost hunters. I was like, "OK, smart guys, this is what's happening to me, and if you have an explanation for me, I would love to hear it." And one of them said, "Have you heard of carbon monoxide poisoning?"

Once the house was checked, it was discovered that the levels of CO were dangerously high. Poppy was told she was in danger of going to sleep and not waking up. What if that anonymous ghost hunter had said, "My group can come investigate in two weeks. Does that work for you?" He would have never investigated because she would most likely have been dead. Instead, he asked a very simple question and quite possibly saved her life.

If there's any greater proof that one's primary duty, as a ghost hunter, is to discover the truth, and not to "prove the existence of the paranormal," I don't know what it might be.

But how does one go about finding the truth? To that end, modern technology now offers a host of options.

Smile For the Camera

Ghost hunters have always been keen on technology. As soon as photography was invented, people began attempting to take pictures of otherworldly things. Of course, with the invention of such trickery as double exposures, there were plenty of people making "ghost pictures" who were simply selling a lie, but that didn't mean that every single ghost photo was a deliberate fake.

Early photography used glass slides or metal as media by which pictures were recorded. Photography was laborious and expensive. People often called a photographer to come take a picture of a relative who had just passed because, if they had no other picture, this was their last chance. Expense suddenly became less of an issue.

Not the Brown Lady. This is a double exposure from 1899. (Wikimedia Commons)

Then newer types of photography, such as tintypes, were invented. Tintypes were quick and cheap. The equipment was portable. Photographers began going from town to town selling their services. But soon, cameras became even more portable and affordable. Everyone and their brother suddenly had a camera and was willing to use it. Photography, and ghost photographs, flourished.

One of the most famous is that of the Brown Lady of Raynham Hall. I doubt there's a need to go into much detail because the people who are likely to be reading this book are also likely familiar with the photo. The main question to be asked—is it real?—is unfortunately, impossible to answer. The magazine for which the picture was taken, *Country Life*, has kept the negative in their offices and refused all requests to examine it. So who's to say if the photographers faked the photo, if a ghost really walked down the stairs in front of them, or if something else happened? Strangely, the Brown Lady has remained unseen since that day she came down the staircase for the photographers of *Country Life*.

The Brown Lady photograph's story is, ultimately, similar to many famous photographs. No matter how appealing the story, there always seems to be something that leads to some skepticism. Was the negative messed with? Did the photographer have ulterior motives? Was the photograph simply a trick of the light? Cameras, no matter how sensitive they are, cannot yet compete with the human eye (though they likely soon will) in terms of sensitivity to light. Just because the eye sees something doesn't mean the camera will be able to accurately capture the same thing.

Modern cameras are much better than they used to be, and of course, we've gone digital, which has its own advantages and disadvantages. Ghosts are showing up so much now that entire YouTube channels are devoted to showing the pictures and videos of the "unexplained."

I watch numerous clips of "real" hauntings caught on camera and uploaded to the internet, always hoping the next one might actually be as scary as advertised or even genuinely creepy. They rarely are. Sometimes, they seem like rather obvious fakes, despite the host of the video doing his or her best to cite the mysterious nature of the clip. Once in a while, something genuinely interesting happens. But I think the bulk of them are often merely compression artifacts.

Files written in a digital format, especially if they are photos or videos, can be exceedingly large. Thus, many formats compress the photo or video to a more manageable size. The trade-off is that the results can be quite grainy or jerky. You don't even have to reach for an explanation like a bug on the camera lens to realize that poor quality video that has been compressed is going to be full of little blips and splotches. The phrase I've heard used for this type of video or still camera is "potato cam," that is, footage so poor it might as well have been shot with a potato.

It doesn't help that anything in the photo that someone doesn't understand qualifies as paranormal. That insect or speck of dust that's reflecting the flash? Spirit energy in the form of an orb! Maybe it's a ghostly hand! Strange mist? Surely it's a ghost and not the operator's breath or a mist or even some dust stirred up by the operator's footsteps.

One type of photo that has been explained is that of *sky fish* or *rods*. These are elongated shapes that show up on digital photos but weren't visible to the photographer at the time. It turns out they are a result of interlacing and thus merely a trick of technology and not something paranormal (or extraterrestrial, as is sometimes claimed). This disappoints me greatly; I was hoping to someday see a sky fish myself. Turns out they only exist inside the camera.

Overwhelmed sensors can make black splotches appear in photos and videos. Light bouncing around inside the camera can make the aperture itself show up in a photo. While these two phenomena are generally mistaken for UFOs, they may also give rise to ghost photos, depending on the beliefs of the photographer.

On one hand, photography is so pervasive these days, it would seem that photography should be one of the best ways for proof of ghosts to be found. Yet the ease with which photos can be altered and the downsides to photography such as compression artifacts means that, perhaps, photography will never be reliable enough for proof of the paranormal to be found that way.

EMF (Electromagnetic Field) Meters

Operating on the theory that, if spirits exist and are energy, they must be operating in some band of the electromagnetic spectrum, EMF meters are promoted as being able to help locate these anomalies. The meter is actually designed to detect things like faulty wiring, but ghost hunters are willing to bank that the meter will tell them where the ghosts are likely hanging out.

The logic *seems* sound, but ultimately, no one has ever been able to explain just what sort of anomalous energy signatures would actually equate to ghost activity. Nor have I seen anyone do an experiment like record the EMF levels in every house in a neighborhood in order to get baseline measurements for the area. How can you know if there are anomalies if you don't even know what's *normal?* Yet EMF meters are cheap to acquire, easy to use, and help both investigators and the people who have asked for their help believe that actual science is being done.

Voice Recorders

Perhaps one of the easiest way to attempt to capture a spirit voice is by using a device to record the sounds of the environment, known as Electronic Voice Phenomena. Generally, one uses a tape recorder or, in the new digital age, a voice recorder app on your phone. One simply opens the app, asks a few questions, and then listens to the playback at a later time. If odd sounds are heard that the operator did not hear when asking the questions, it is assumed these noises are paranormal. They might even turn out to be words, though other sounds, such as growls or moans, have also been recorded.

If cameras really aren't going to pick up anything in the environment that the human eye can't see, then digital (or even analog) recorders are also unlikely to be able to pick up anything the human ear can't hear. EVP recordings are often highly filtered and, even then, the supposed spirit messages are still unintelligible. However, if the ghost hunter thinks he or she knows what the spirit is trying to say, they will consider their attempt to communicate with the spirit(s) a success.

Spirit Boxes

If you're too impatient to record and then listen to a playback at a later time, you can have some immediate feedback by using something like a spirit box. The spirit box is a piece of equipment that scans the AM/FM bands in rapid succession on the theory that spirits can use this device to communicate using words they choose from various radio broadcasts.

If words do indeed come out of the spirit box, they usually have no obvious meaning. The ghost hunters using the device must decide upon the context and what the word might mean.

Who Ya Gonna Call?

Interestingly, ghosts often seem to use a device that has been part of our everyday lives for a hundred years: the telephone. Almost from the very beginning, people reported odd voices or sounds when connections were made.

According to researchers Hall and Petonito, phone calls from the dead have been increasing in frequency. One theory says that this

> I think what makes phantom messages so baffling is that we don't understand them, where they come from, how they are generated, and the motivational force behind them. We always seem to be drawn to that which we do not understand and cannot replicate upon command.
> Laurene M. Gomez, Psychologist

is due to the increased connectivity in our world. We don't live in a world with a few cell phone towers or electrical substations; we live in a world with millions of cell towers and substations. As more devices become wireless, why wouldn't ghosts be able to tap into all that electromagnetic radiation and use it to deliver messages to the living?

On the other hand, apparently, phone calls from charged/plugged-in devices happen at about the same frequency as those coming from dead battery/not plugged-in devices, so it would seem ghosts don't necessarily even need the device to be functional to use them.

Messages from the dead sometimes consist of strange voices repeating a word like "hello." Or the message is in the voice of a loved one saying something like "Good-bye," or "I love you." Sometimes, the voice lets the living person know "I have to go" or it will ask for help. The voices rarely give too much detail about what they are experiencing or what the afterlife might be like. Hall and Petonito report that at least one phantom caller left the message "The lightning is beautiful here." But, of course, the voice does not explain where "here" is.

Sometimes, the voices can be abusive, shouting terrible messages like "Die, bitch." Just who these angry voices are is something the recipient of the call generally does not recognize the voice. If one attempts to dial the number back (assuming a number shows up at all), the call will not go through.

Besides phones, messages have come via email, over cable boxes, radios, and televisions. Those who have the wherewithal to get their messages through appear to be able to use any electronic device available. We can, I assume, expect that as we gain more and different and more powerful devices, the messages will continue.

<u>Fun With Friends</u>
So, what are we to make of these messages? Are they really from the dead?

Sadly, our answer has to be, *who knows?* Just like human brains are imperfect, and cameras have their issues, electronic devices such as phones and televisions also have issues. These issues can even be caused by animals such as rats, mice, or other vermin. One family's security system began going off at random times three days after the matriarch of the family died. The family felt she was trying to reach out to them. However, once the security system company sent out a repairman, he discovered that roaches had infested the system and were tripping the connections. Once the roaches were evicted, the system worked as it should.

Perhaps grandma was really trying to communicate and the roaches were just the most convenient way for her to do that. Can we really say that's not what happened?

The other thing to consider is that ghost hunters are basically ordinary folks who are intrigued by the idea of the paranormal, but do not necessarily have a good handle on how to do actual science. As Benjamin Radford points out in his article, "The Shadow Science of Ghost Hunting," "Just as using a calculator doesn't make you a mathematician, using a scientific instrument doesn't make you a scientist...you may own the world's most sophisticated thermometer, but if you are using it as a barometer, your measurements are worthless."

Radford does his own research into the paranormal using low-tech non-flashy gear like notebooks, flashlight, and a tape measure. Meanwhile, the other ghost hunters he's with have their scientific equipment, which they may not know much about. When Radford asked a fellow ghost hunter why he used an EMF meter, he was told that, "At a haunted location, strong, erratic fluctuating EMFs are commonly found. It seems these energy fields have some definite connection to the presence of ghosts. The exact nature of that connection is still a mystery." Ultimately, the other ghost hunter admitted that there was no way to substantiate any of this.

Radford says, "Of course, no one has ever shown that any of this equipment actually detects ghosts...If a device could reliably determine the presence of absence of ghosts, then by definition, ghosts would be proven to exist." Unfortunately, his critique of many ghost hunters is all too on point. How often in ghost hunting TV shows has the viewer been treated to an "anomaly," whether that is an area that is a few degrees cooler than its surroundings, or an EMF meter detecting an EM field, or someone feeling creeped out? This will be cause for the host of the show to declare they have found something paranormal. "Because the standard of evidence is so low, it's little wonder that ghost hunters often find "evidence" (but never proof) of ghosts."

Radford concludes:

> If ghosts exist, why are we no closer to finding out what they really are, after so much research? The evidence for ghosts is no better today than it was a year ago, a decade ago, a century ago. Ultimately, ghost hunting is not about the evidence (if it was, the search would have been abandoned long ago). Instead, it's about having fun with friends, telling ghosts stories, and the enjoyment of pretending you are searching the edge of the unknown. Ghost hunters may be spinning their wheels, but at least they are enjoying the ride.

And yet, as Carrie Poppy says, perhaps we shouldn't completely abandon all hope that actual evidence may some day be found.

Maybe I'm just eternally optimistic, but I hope I never lose this hope, and I invite you to take this same attitude...I still get excited at ghost stories every single time. I still consider that every group I join might be right, and I hope I never lose that hope.

So, yes, we can still hope. But sometimes, roaches are just roaches.

Is Anybody Out There? Do You Care?

Rather than love, than money, than fame, give me truth.
Henry David Thoreau

We live in a fantasy world, a world of illusion. The great task in life is to find reality.
Iris Murdoch, novelist

***

Perhaps you're thinking, why is any of this important? Who cares if people believe in ghosts if they don't exist? Isn't it just fun to watch the television shows and watch the ghost hunters freak themselves out in the dark? What harm can a belief in the afterlife do?

Well to start off, let's recall Carrie Poppy's story about a haunting that turned out to be carbon monoxide poisoning. But she only found out about the carbon monoxide *after* she and a friend had tried to dispel the ghost by burning sage. If she'd skipped the smudging and gone straight for checking carbon monoxide levels, she would have not only gotten rid of her ghost far earlier, she'd have breathed a lot easier (literally).

A disbelief in ghosts could, in some instances, literally save your life.

But honestly, how often is a ghost caused by carbon monoxide? Probably not that often, certainly not often enough to account for every ghost encounter. A belief (or even a willingness to believe, which is a bit different) in the paranormal can have interesting positive and negative consequences.

Ghosts and Disaster

After a natural disaster, many resorts and other tourists destinations have had to rebuild their businesses. Marketing companies have discovered that the most important factors in getting the tourists back include rebuilding the area to pre-disaster levels and a marketing message that puts the location in front of people who may be booking their next vacation. The recovery of the area can be negatively affected by such things as issues with insurance, ownership of properties, the level of physical damage, or other things which can keep locations closed for an extended period of time. To overcome any negative issues, resorts and other tourist-dependent businesses may offer incentives such as discounts or extras to encourage people to come back. However, research has shown that such enticements do not work if the potential tourists perceive the destination as being unsafe for whatever reason.

What I was surprised by was that ghosts are sometimes one of those reasons. In areas affected by the 2004 Indonesian tsunami, tourism was down for some time. Of course, it was depressed by the need to rebuild infrastructure, but also, as Bongkosh Rittichainuwat says:

> (M)any inbound Chinese and Thai tourists substituted their original travel itineraries to tsunami-affected areas with trips to other beach resorts due to perceived risks affiliated with ghosts...In contrast, it is almost a given that Westerners would not consider the issue of ghosts important in deciding whether to visit such places.

Though many Chinese and Thai people are Buddhists, Hindus, or Taoists, and these religions feature a belief in the transmigration of the soul, which states that souls leave a body only to re-enter a new one in the cycle of rebirth, participating in such social rituals like ancestor worship and having a cultural predilections for telling ghosts stories mean people do not necessarily strictly follow a belief in transmigration. Instead, many people believe that souls can wander the earth for some time after death.

Adopting a belief in ghosts from one's elders is common.

Thai children are taught about and are made well aware of the existence of ghosts and their potential dangers, and are taught to chant protective spells and to wear small Buddha amulets around their necks to ward off ghosts. Likewise, Chinese children are taught to avoid proximity to dead bodies as they are believed to bring bad luck. The fear of ghosts is instilled in children very early and continues into adult life.

Especially dangerous are the souls of those who die before their time. Since they were unable to live to the end of their natural lifespan, they are jealous of the life that the living have which they had ripped from them. These ghosts become malevolent.

The 2004 tsunami killed over 225,000 people. That's a lot of people who died before their proper time, and therefore a lot of spirits wandering those beautiful Thai beaches (among other places) being malevolent and looking for living people to attack. Tourists from China, Taiwan, Hong Kong, and Singapore avoided places ravaged by the tsunami because they might bring a ghost illness back home with them. Fever, difficulty breathing, even mental disorders can be caused by ghosts. It only makes sense, if ghosts can cause such things, to avoid places where a lot of newly-minted and malevolent ghosts are suddenly going to be hanging out.

Additionally, holidays proved a difficult time to get toursts back, as respondents reported a belief that encountering ghosts during New Year's celebrations would bring bad luck throughout the entire year.

By the time five years had passed, many Asian tourists had overcome any concerns about malevolent ghosts and were returning to those Thai beaches that had previously been affected by a belief in ghosts. This was partially due to a belief that crowds could somehow dispel the malevolent ghosts (or at least convince them to pack up and leave), as well as believing that certain religious ceremonies may have been performed which could have altered the ghosts from malevolent to benevolent. Once benevolent, the souls would then be eligible for rebirth and would no longer trouble the living.

Pandemics & Quarantines

A strange wrinkle of the current pandemic and subsequent quarantining in our homes, is that some people have discovered they are not as alone as they think they are.

Before the pandemic, people left their homes for many reasons: to go to work, to buy food, to visit friends and family. Once lockdowns became the norm, people found themselves at home for day after day without a break. Perhaps not surprisingly, some of those homebound people began to wonder what was up in their houses and apartments. Molly Fitzpatrick reported on people who suddenly discovered that they had doorknobs that rattled by themselves, window shades that banged against the window frames for no reason, and the sounds of footsteps in attics or in the empty apartments above them. One person said he was so freaked out by the banging window shades that he thought he was caught in an earthquake. "I very seriously hid myself under the comforter...because it really did freak me out."

One person reported that, after getting up in the middle of the night for a glass of water, he spotted a WWII era soldier sitting at his table. When he did a double-take to get a better look, the ghost had disappeared.

A paranormal researcher interviewed by Fitzpatrick said that, before the lockdown, he'd get two to five reports a month, but after the lockdown, he started getting five to ten per week. He reported that he'd experienced a similar spike in calls right before Y2K. "It does seem to have something to do with our heightened state of anxiety," he said.

Ultimately, the likely culprit for many of these new hauntings is that people aren't used to being home all the time. Even on the days they weren't at work, they may have been out getting their errands run, or perhaps they'd have friends over. But those activities dried up in the pandemic along with the off-site work. Now those noises come to the forefront.

It's not just noises, either. Some people report increased incidences of feeling as if they're being watched or finding personal items on the floor that had previously been hung up or sitting on a table. Sometimes, objects like keys disappear entirely or are moved to different locations, while in one case, a person reported that a camera lens that had been lost for years suddenly showed up on her nightstand.

Not every newly-noticed ghost is unfriendly. One person reported that their ghost seems determined to smooth out their bedsheets; he interprets this as the ghost trying to make him "feel more comfortable."

Ultimately, the anxiety that has come with the pandemic may be causing hauntings by making people hyper-vigilant to their surroundings, or helping them feel less alone by granting them a new companion. Perhaps it's that people are rearranging their mental furniture and schedules along with their physical lives. "Maybe you're listening more closely in the greater quiet all around us."

Even if you are, is that strange sound you hear in the evenings just your brain helping you cope with isolation, or is it really a ghost you've simply been too busy and loud to notice before? Only you can know.

Going Viral

In his book, *Contagious: Why Things Catch On*, Jonah Berger notes that most things which go viral do so for one (or more) of six reasons:

Social Currency: Does knowing a thing give you a certain cache that you can't get any other way with your friends and coworkers? Do you know the secret code to get a coupon at a local bar? Do you know some fact that everyone thinks is cool but you're the one who's spreading the news about it? You're the one with the inner knowledge that makes you a member of the cool kids' club.

Triggers: One of the most infamous examples is the Rebecca Black song "Friday." Though it was excoriated for being a terrible song, guess what day of the week constituted the day the song was downloaded far more than any other? If you guessed anything other than Friday, I suspect you are merely being sarcastic.

Emotion: Engage people's emotions and you'll have them eating out of your hand. Memes or posts on Facebook that successfully make people feel something are liable to have dozens, if not hundreds, of comments reading, "I'm not crying, you're crying" or "Who's cooking onions in here?" to show that they were affected by what they'd seen and/or read.

Public: people are more likely to have positive associations with something if they see others using them in public. For this reason, the folks at Apple debated how to place the Apple logo on their laptops. They decided to make it right side up from the point of view of someone across the room seeing the product being used. After all, the person using the product had already purchased it, and Apple knew they then had to reach the other people around the owner to increase sales.

Practical Value: Do you have a life hack that can help people organize their kitchen? Or a way to store their earrings? The one I've used ever since reading about it is using old buttons to keep my earring pairs together. Works like a charm (assuming you put the earrings back on the button once you take them out of your ears)! Give people a practical way to make their lives easier and watch your shares, retweets, or likes go up.

Stories: People like to tell stories. You can hardly hear someone talk about finding a new hairdresser or a new television show without also hearing about how they found it, who they found it with, and how it's affected their life. Usually, people love to hear these details. Pet product ads are constantly telling you, "Now that I feed my pet this new food Brand X, they're happier and healthier than ever." You want your pet to be happy too, right? Then surely you'll give this new brand a try!

When it comes to stories, though, I'd like to carve out an exception for those recipe sites online where you have to scroll for about a thousand pages of "my experiences with this recipe and all the ingredients plus meeting the cow the butter came from" before getting to the actual recipe. Those are stories I could do without. Just give me the freaking recipe already!

Hauntings hit several of these sweet spots. Are you the person who knows the *real* story of the haunted house down the street? That combines social currency with stories and even emotion. Three reasons for something to go viral wrapped up in one. So how can we be surprised when ghost hunting shows, ghost tours, and even equipment that promises to help you find ghosts are popular in all seasons and times?

I suspect it's these reasons that keep hauntings later shown to be hoaxes in the public eye. These are stories that drill their way into the human mind for reasons such as the storytelling or the emotions they trigger, and thus will never be eliminated from public consciousness.

Personally, I would prefer for those hauntings which prove to be hoaxes to promptly be dropped by everyone, so that I (and the rest of us) can focus on the ones that appear to be truly mysterious. If we're ever going to have proof of anything bizarre or paranormal (this goes for cryptids, UFOs, ghosts, and every other out-of-the-ordinary thing), then we need to be focusing on the unexplainable, and not just the ones that make good stories.

The Amityville Horror is a case in point. The book by Jay Anson contains "facts" which cannot be corroborated, and others which are simply common tropes, such as the "fact" that the house was built on an "old Indian burial ground." What haunted house worth its salt would agree to be built anywhere else?

The Amityville Historical Society claims that the "facts" in the book about the local area aren't true, and the tribe Anson claimed lived in the area actually lived elsewhere. The local police say they were never called to the house as Anson claims in the book. In fact, the entire saga gets shadier the more one looks into it. Weather reports dispute the weather on the dates given in the book. For instance, there was no snow on the ground on the day George Lutz claimed to have found cloven hoof prints in the snow. Reportedly, Anson told a fellow writer that "one day I predict that you are going to be sitting there broke, writing your little non-ghost stories, and I am going to be on an island out in the Bahamas..."

Perhaps Anson didn't say that. Perhaps George Lutz merely misremembered the day he saw the cloven hoof prints in the snow. Perhaps the police are covering up all the times they were called out to the house. Perhaps the local priest is part of that cover-up. Maybe the local historical society has a grudge and so is trying to make Anson and the Lutzes look bad. I mean, I guess everyone *not* making money from the book and movies could be involved in a far-reaching conspiracy, but honestly, that's even harder to buy than a demon pig trotting around an Amityville house in the non-existent snow.

Ultimately, the important point here isn't that the story is more than likely a complete bunch of hogwash. What we need to remember is that this story is now part of the public consciousness and clearly isn't going anywhere. Once something's gone viral, it has staying power. I disagree with Robert Bartholomew and Joe Nickell who refer to the "stubborn persistence" of this modern myth as "remarkable." It is, actually, *exactly what we should expect* from a story that's engaged people's emotions, splattered the culture with triggers, told a good story, and given some people the social currency of being "in the know" about "the truth" of the situation. How can mere skepticism defeat such?

Bartholomew and Nickell conclude, "The myth has taken on a life of its own, disrupting the everyday existence of many of the residents in this small seaside village. It is a story that is all too familiar to skeptics— exploitation by elements of the film, documentary, and publishing industries, eager to profit from human vulnerability. That is the real horror of what happened in Amityville."

Animals and Their Souls

We patronize the animals for their incompleteness, for their tragic fate of having taken form so far below ourselves. And therein we err, and greatly err. For the animal shall not be measured by man. They are not brethren, they are no underlings: they are other Nations, caught with ourselves in the net of life and time, fellow prisoners of the splendor and travail of the earth.
Henry Beston, author

If having a soul means being able to feel love and loyalty and gratitude, then animals are better off than a lot of humans.
James Herriot, author

You think those dogs will not be in heaven! I tell you they will be there long before any of us.
Robert Louis Stevenson

✳︎✳︎

There's a story I've read and heard from multiple sources that goes like this:

A man had just died. The moment he arrived in the afterlife, he found his long-dead faithful dog waiting for him and was overjoyed to be reunited. The two of them walked down a road until they came to a beautiful marble gate with a stately man sitting beside it.

"Is this heaven?" asked the dead man.

"Yes, it is. Come on in!" said the gatekeeper.

The man and his dog headed for the gate, but the gatekeeper put up his hand. "He'll have to stay out here," he said. "No dogs allowed."

The dead man was saddened to have to miss out on the joys of heaven, but there was no way he would leave his dog now that they were finally together again. He turned away and continued on down the road. Some ways later, he saw a man sitting by a run-down shack.

"Can I go in?" asked the dead man.

"You're more than welcome," he was told.

"And what about my friend here?"

"We love dogs. He's welcome, too!"

"So, is this heaven?" asked the dead man.

"Yes," said the man.

"But the gatekeeper at the last place said that was heaven."

"Yes," said the man again. "We don't mind that they say that. We figure anyone willing to leave their dog behind isn't anyone we want, anyway."

The question of whether or not animal have souls has been on many people's minds for thousands of years. If you believe in the transmigration of souls, then obviously animals have souls. If you believe in a Christian heaven, then the question of animal souls is much more fraught with theological issues. The Bible appears to have no opinion on the matter.

Today, the issue is discussed in books such as *Signs of Pets in the Afterlife*; *The Amazing Afterlife of Animals*; *Will I See Fido in Heaven?*; and *The Pet Soul Book*. Clearly, many people have come down on the side of answering the question about animal souls with "yes." It can be quite painful to lose an animal companion. How could an animal who had such a deep bond with a human *not* have a soul?

Accordingly, people have been spotting animal spirits for some time. Dogs are probably the most commonly reported animal spirit, but cats have been spotted as well. Phantom horses, often pulling phantom carriages, are certainly seen in some towns.

A demon cat has been seen in the U.S. Capitol, though apparently it hasn't been spotted for some time. One story has the cat starting out cat-sized but growing to the size of a tiger by the time it got to the poor late-night watchman unfortunate enough to see it. A current explanation for the demon cat is that the cat was most often seen by men who, more than likely, were drunk. Once the security forces at the Capitol were expected to show more professionalism, sightings of the cat dropped to zero.

Ghost dogs include rat terriers chasing down equally spectral rats, late canine pets announcing their presence by scratching at doors or whining for food. Some have been spotted checking the floor under the table for scraps. Others have seen ghostly dogs wandering through their homes, though the dogs were not any dog they had owned at any time. These ghost dogs seem to haunt very old houses, and the current owners attribute the strange sighting to a dog that must have been owned by a previous tenant of the property. I'm not sure anyone has ever managed to substantiate that feeling by researching the kind of dogs previous owners had.

Perhaps the most common way for ghost dogs to let the living know of their presence is by barking. This sounds suspiciously like they might be living dogs being mistaken for dead ones, but who knows? Perhaps people's canine companions want to reassure their owners they are still around and still looking after the humans they had to leave behind.

Two white phantom horses pulling a phantom carriage have caused street accidents in Hollywood. Four black horses pull a phantom carriage near Santa Ynez, California, though this is sometimes considered to be less a case of ghostly horses and more demonic ones, as the owner of the stagecoach is said to be the Devil.

A phantom coach was seen in Surrey, England, in 2007. The driver who spotted it reported that it crossed the road in front of him even though there was no crossroad to be seen. In Northamptonshire, a phantom coach pulled by phantom horses drives up and down the lane where the driver ran over his daughter's lover centuries ago.

Animal ghosts have been experienced in non-Western cultures as well. A Hong Kong teenager related how she lived near a site where a slaughterhouse used to stand. When the slaughterhouse had been in operation, she had heard the pigs squealing as they were killed; "their voices carried far away. I always felt sorry for the pigs and shed tears for them." After the place closed and was torn down, she still heard the pigs.

"We could hear pigs crying like when they were being slaughtered. My hair stood on end when I heard it."

Even the Ghost Hunters Have Pets Return

Raymond Bayless, who was a parapsychologist as well as an artist whose work has hung in the National Air and Space Museum, and his wife Marjorie, reported being haunted (briefly) by their cats after the cats' deaths. In each case, their surviving cats did not react to the return of their companions. However, these hauntings generally consisted of Bayless or his wife hearing a soft meow or catching a glimpse of what appeared to be their former feline friend out of the corner of their eyes. Both Baylesses claimed to be familiar enough with hauntings that they felt they could not be tricked by their senses (Bayless was the author of several books on parapsychology, including 1970's *Animal Ghosts*), but it's far too easy to see that anyone, even a parapsychologist, who was grieving for their beloved pet, might hear or see something odd at least once or twice in the days after their pet's passing.

Animal Ghost Stories Should Get More Play

D. Scott Rogo, in his book *The Haunted House Handbook*, has an entire chapter on animal ghosts (it's where I found the story about the Baylesses cats), but his chapter is barely longer than mine, and he doesn't even mention ghostly horses. Despite the fact people have been reporting animal ghosts for some time, the number of stories out there seem to be quite limited. Perhaps I'll find more stories out there and learn more about the kind of animals people see, and in what circumstances.

Strangely, outside of the stories from Rogo's book, the stories I can find, like the ghostly carriage-pulling horses, sound more like local legends than anything more substantial. And let's face it, stories of the Devil coming to Earth to find souls to steal is hardly a new trope. That his coach and his horses are black is just part of the standard tale.

If anything, the Hollywood coach that causes traffic accidents is the most interesting, since something that has that effect on the physical world would seem to be worth investigating further. Perhaps the coach with its white horses is merely a reflection of something that shows up at a certain time of day when the sun is at just the right angle. Or perhaps the drivers are drunk and know the story, and so use it to distract from their inebriated state. Or maybe a phantom carriage really does speed down Lookout Mountain Road in Hollywood, silent and frightening, and disturbing enough to make people wreck their cars. Who can say for sure?

So I Looked...

The skeptic does not mean him who doubts, but him who investigates or researches, as opposed to him who asserts and thinks that he has found.
Miguel de Unamuno, author

I'm a real open-minded skeptic when it comes to the ghost world. And there are places that I've traveled to where it's very scary and I end up feeling like there might be something going on there.
Josh Gates, television producer

I think anybody who's curious about anything, including their own mind, is inherently a skeptic.
Jamie Hyneman, MythBuster

Your brain is constantly choosing what it believes to be the most significant aspects of your surroundings and paying very little attention to anything else.
Richard Wiseman, *Paranormality*

*
**

So, yeah, I don't believe in ghosts. I've never had an experience that was unexplainable, and I've never heard a story that convinced me the paranormal is definitely real.

On the other hand, plenty of people do believe. Their stories are funny, wrenching, thought-provoking, and intriguing.

Wiseman quotes studies that say roughly 30% of people believe in ghosts, and 15% claim to have experienced them. Dagnall and Drinkwater found similar figures, with 38% of people saying they believed in ghosts. A Harris poll from 2013 showed that 42% of Americans and 52% of the British believed in ghosts. So, take whichever figure you find most believable. Quite possibly, all these figures are true, as it is unlikely that the pollsters in each case phrased their questions in exactly the same way, and *how* questions are asked in polls is at least as important as *what* is being asked.

So, it should be more or less accurate that between one-third and one-half of British and Americans believe in ghosts. Of course, that means that up to two-thirds don't.

Even those who don't believe have their creepy stories. Tiffanie Wen talks about her own experience in her article "Why Do People Believe in Ghosts?" She had taken some photos of her apartment with her iPhone. A few weeks later, she was scrolling through the photos when she discovered that one of them contained the image of a man. The man was no one she knew, and she had been alone in her apartment at the time the photo was taken. So, who was the man? Why was he in her photo? Wen has no answers, but she does say, "I've been using the photo to scare my friends, and myself, ever since."

Wen's friends came forward with their own stories. While in Ethiopia, one had used her iPhone to take a photo of a boy looking at his hands. When she looked at the photo later, a second boy was in the photo, and this boy was looking straight at the camera. Considering both these photos showed up on iPhones, Wen contacted Apple for a comment about their product and the paranormal; an Apple spokesperson looked at the images from Wen's and her friend's iPhones but had no comment. Still, since these images were captured by the same Apple product, the possibility that these images are due to some glitch in the camera software has to remain one of the front-running explanations.

So there's certainly nothing stopping me from investigating the tales of hauntings and hoping for an experience of my own. To that end, I have stayed in haunted hotels and B&Bs, attempted to get EVPs in graveyards, and have taken photographs in allegedly haunted spots where odd things show up in pictures. Here's a bit of a rundown of haunted spots where I've done a bit of sniffing around.

Hannibal, Missouri: Garden House Bed & Breakfast

We stayed here with some friends and since my friend has been staying in haunted venues for years, I thought she should take the "most haunted" room. Her companion did possibly feel something touch him when he woke up, but of course, having an anomalous feeling or sighting upon waking is fairly common and doesn't necessarily indicate something paranormal. Neither my husband nor I experienced anything unusual. However, this location is a very nice B&B and is one of the stops on the Hannibal ghost tour.

Virginia City, Nevada

Like many towns in Nevada, Virginia City was founded upon the discovery of silver in the area; in fact, it was the site of the famous Comstock Lode. It eventually turned into a town of 25,000 people. The boom resulted in an explosion of businesses, overnight millionaires, and the establishment of newspapers, one of which employed a then 33-year-old Mark Twain.

What booms eventually busts, and Virginia City, over a century past its heyday, is now a quaint historical town with a fair number of ghostly legends and haunted buildings. We stayed at the Silver Queen Hotel, which allegedly hosts a multitude of ghosts, including two that have been named Annie and Rosie. These two apparently like to run through the hallways and tap on windows. The Silver Queen is a nice place to stay if you can manage stairs (you have to climb 29 of them just to get from the street to the hotel level above the bar).

The closest experience we had that could even, by a great stretch, count as slightly out-of-the-ordinary was an odd knocking on the wall behind me while I sat in a chair. However, there is a private residence on the other side of that wall, so I wouldn't discount the theory that these were merely the sounds of an all-too-human and non-paranormal neighbor. Or maybe pipes or something else in the shared wall. But who knows?

Tonopah

By the time the year 1900 rolled around, the Comstock Lode was a distant memory. Nevada was experiencing financial difficulties with its boom days apparently long behind it. However, that was when Jim Butler discovered silver near Tonopah Springs. Within a few years, the town gained a multitude of new businesses, schools, and banks. Eventually, hotels sprang up, including the Mizpah Hotel, which, for a time, was the tallest building in Nevada.

We stayed in two different allegedly haunted locations in Tonopah.

Mizpah Hotel

The most famous ghost in the hotel is the "Lady in Red." You can book

the room where she is seen the most often —it's on the fifth floor—but, fair warning, it's more expensive than other rooms. Since the Lady in Red is also seen elsewhere on the premises, we did not put down the money to stay in the Lady's favorite room.

Spirits have been known to get in the elevator with guests, so we took some pictures there. Some people have reported music playing on the elevator when no music is being piped in there at all, and a few people have heard a voice tell them to "get out." As far as I can tell, we did not capture anything unusual in our photos, and we did not hear anything unusual. The most odd thing that happened to us was that, on one of our elevator journeys, we were unable to get the button for our floor to light up (floor 4). The people who came into the elevator behind us pushed the button for their floor (floor 5) and once the elevator delivered those people to their floor, we were able to press the button for 4 and have it light up. The elevator descended and allowed us out on 4. Were we being pranked by a ghost who wanted to prioritize guests on the fifth floor over those on the fourth, or was it just the wiring in an old building?

What do you think?

The Clown Motel
Ever since I saw a travel show that featured this location, I've wanted to stay there. Imagine: a motel full of clowns! Not only is there clown art on every room door, but every room has clown art on the walls; the office features a clown-themed

Clown Motel next door to the Old Tonopah Cemetery

gift shop; and there is a clown museum (really, more like a display of lots of clowns of all kinds).

I was pretty neutral about clowns before but now I am much more of a fan of clown iconography. Anyway, the motel is considered haunted, perhaps because the Old Tonopah Cemetery is in the next lot over.

We noticed a ghost hunting group at the motel on the last night of our stay. So investigations of this location are ongoing by various interested parties.

The Old Tonopah Cemetery
This cemetery was used for only about ten years, and was closed when mine tailings kept damaging headstones. A new cemetery was opened and is still accepting burials today. You can download a walking tour of the old cemetery from

www.tonopahnevada.com/CemeteryBrochureOnline.pdf.
We conducted a couple of EVP sessions in the cemetery but I don't think we found anything. You can listen to our recordings on my YouTube site's playlist "Escaping Normal."

Tonopah Stargazing Park
I list this site because I conducted a short EVP session here while Todd was setting up his cameras to try to get photos of the night sky. Tonopah is noted for having some of the clearest and darkest skies in the country, so if you want to take photos of the stars, this is the place to go. However, while we were there, wildfires in California were sending up so much smoke that the sky wasn't clear enough to get good shots. This EVP session is also uploaded to my YouTube playlist "Escaping Normal."

Goldfield Hotel & Goldfield Cemetery
Goldfield is another mining town that is only about twenty miles from Tonopah. The Goldfield Hotel is not open as a hotel, though ghost hunting groups are sometimes allowed in to conduct their investigations. We were only able to observe it from the outside. Photos of the lobby were taken through the windows.

The Goldfield Cemetery is also allegedly haunted. We visited in the late morning after we'd gone hunting for chalcedony at the local gem field. As you should suspect by now, even if the paranormal is out there, it does not seem eager to show itself to us. We did not experience anything unusual in the cemetery.

El Paso

Concordia Cemetery is well-known to be haunted. We joined a ghost tour of the place on a pleasant October evening. While neither my husband nor I felt, heard, or saw anything unusual, our friend Brian did feel as if he were being observed. He has experienced this before in the

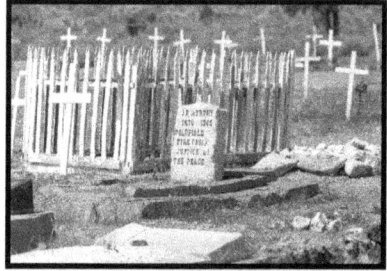

Goldfield Cemetery

cemetery. He has also experienced a feeling of malevolence at one location, which pushed him back across the path.

Many of the otherworldly experiences in the cemetery sound like general haunting stories: the sound of the giggling of children in the children's section; shadowy figures showing up in photographs; people getting odd sensations or feeling creeped out. Still, whether or not you think something paranormal is going on in the cemetery, it is an interesting place to visit. Allegedly, over 60,000 are buried here, so this city of the dead is quite large. Who knows who you might meet within its gates?

Brian shared his experiences with me on camera and you can watch that on my YouTube "Escaping Normal" playlist. For more from me and Brian, you can view our "Infernal Intercourse" playlist, which is exclusively about the TV show *Lucifer*. Brian also joined me and Steve Hill for an episode on my "Satanist Says What?" playlist.

You can find my YouTube channel by searching for Marella Sands once on YouTube.

The mausoleum in Concordia Cemetery that felt threatening to Brian. This (the south side) is supposedly the side where the greatest number of experiences happen. Author on the far right.

The Great Hope

There is no conclusive evidence of life after death, but there is no evidence of any sort against it. Soon enough you will know, so why fret about it?
Robert A. Heinlein, author

Six weeks after his death my father appeared to me in a dream. It was an unforgettable experience, and it forced me for the first time to think about life after death.
Carl Jung

Everyone must leave something behind when he dies, my grandfather said. A child or a book or a painting or a house or a wall built or a pair of shoes made. Or a garden planted. Something your hand touches some way so your soul has somewhere to go when you die, and when people look at that tree or that flower you planted, you're there.
Ray Bradbury, author

*
**

The mere fact that ghosts exist (at least as folklore) only means that humans are still asking that question *what happens to me when I die?* We can see the beginnings of that question 400,000 years ago with those *H. heidelbergensis* individuals who placed some of their dead down that dark shaft in a cave. We see it in Shanidar. And Stonehenge. And any of a hundred thousand other sites where humans and their predecessors have, in some way, done something special with their dead that helps them cope with the overwhelming reality of death.

Let's face it: death is scary. It's not comforting to think of one's consciousness fading into oblivion, is it? It's just as disturbing to think of one's parents or siblings or children as not existing any more. Or what about your faithful pet who shared years of your life with you? Do you want to think about the pet being "dead and gone?" Isn't the Rainbow Bridge much more comforting?

Oblivion may be even more frightening than considering a wretched afterlife where one might suffer or face terrible tests. At least, even if you're in an unpleasant place, you're still *you*. You may be able to escape the torment (depending on your personal beliefs and what sort of afterlife you've been sent to). You may be able to once again be in the presence of loved ones that you've missed terribly. Your faithful pet companions wait at that beautiful bridge and will run to you as soon as you draw near.

Honestly, I love the idea of seeing my pets again. I've had a lot of pets, too, so the meadow where all my late animals are hanging out waiting for my slowpoke self to arrive should be quite large. I can think of nothing better than breathing my last and suddenly finding myself among all my animal companions.

I think I'd be even more thrilled to see my pets than my human loved ones for the simple reason that, for a lot of them, I had to make the decision to "pull the plug," so to speak. I was *responsible* for the timing of their deaths. Sure, they were old and sick and probably hurting, but isn't it so much nicer to think I sent them to a flowery meadow under a brilliant rainbow than to think they simply *ended*?

I admit, though, that even the pleasant thought of the Rainbow Bridge bothers me some. How do I know my pets even *want* to be in a flower meadow? Maybe, if they could be given a choice and understood it, they'd choose oblivion. Just because my human self thinks seeing them in an afterlife would be a great option doesn't mean my non-human companions would agree. Even more than a cultural bias, that's an entire species bias! I mean, even other humans don't necessarily want what I want in an afterlife; how can I assume a member of *another species* would want what I want? I can't.

In the End

As Benjamin Radford pointed out, we don't have any good scientific evidence for ghosts existing, and humans are far too suggestible and easily mistaken to be credible witnesses. Charles Emmons relates the tale of a professor he spoke with in Hong Kong who became the focus of a campus ghost tale just to see if he could. The professor had heard there was an urban legend about a student who had committed suicide after failing a test. He concocted a story where *he* was the professor who had administered the test, added a few details that sounded plausible, and told the story. Interestingly, every time he told the story, he always made sure to relate that *he had made the story up*.

Needless to say, by the time Emmons heard the story, it was being promoted as true, and he was even warned against speaking to the professor as people thought it would upset him to know one of his students had committed suicide due to one of his tests. Emmons (obviously) spoke to the man, anyway, and discovered the truth to the tale.

Here we have a case where people have taken a story and run with it, even when the author of the story freely admits it is fiction. Because we know a bit about why things go viral, we can assume this story had "legs" due to several factors: it was a good story, it gave those "in the know" some social currency, and it engaged the emotions of the students. No doubt many of the students on that campus felt stressed and anxious about tests, and a story about a fellow student who couldn't take the stress and committed suicide was going to be something many would take to heart.

Keep Up the Good Hope

The lack of evidence for ghosts is troubling but people keep looking. And I hope people continue to do so. I still love ghost stories and watching ghost hunter shows. I still love reading the articles and books where people's stories are documented. It's still exciting and wonderful and so so interesting. I'd prefer it if the people doing the looking had a better grasp on science and psychology, but maybe we'll get that someday.

If you're a ghost hunter, I hope you learn to be more skeptical and to engage the subject with more objectivity. Honestly, I once heard a ghost hunter say (I paraphrase; it's been a few years): "We heard footsteps coming from above us. We knew that apartment was vacant, so we knew there was a ghost up there." I mean, *two seconds* of thought went into that conclusion, if that much. Noises + no tenant above us = ghost. It seems to me there could easily have been a dozen other explanations more readily available.

If you're a skeptic, I hope you listen to the stories with an open ear. Maybe, just maybe, you will hear the story that breaks the case for ghosts wide open. Forget the Brown Lady. Forget Amityville. Forget the Borley Rectory, the Winchester House, and any other famous place that has its own ghost stories. Find that story that actually sends a chill down your spine and makes you think, *could this be the one?*

The most convincing ghost story I ever heard was from a friend of mine. She had gone into the bathroom of her Victorian-era home and found a bar of soap in the middle of the floor. The drawer where she kept soap bars was open, and so she figured the dog must have done it. She replaced the soap in the drawer, closed the drawer, and thought no more of it. A short while later, she went by the bathroom again. The drawer was still closed. The dog was (and had been) elsewhere in the house. The soap was back on the floor.

Now *that* is a story I have no explanation for. It might not be flashy, or famous, or come with supporting photos or EVP recordings. But it remains, to this day, the one story I've heard that has any chance of making a believer out of me. It's small. It's subtle. It's difficult to explain.

So, soap-moving ghost, if you're out there, come on by my place. You know where I keep the soaps. I'm waiting.

Works Consulted for the Escaping Normal Series

--, Kate. Personal interview. 12 February 2021.

--, Laura. Personal interview. 16 February 2021.

"1561 Celestial Phenomenon Over Nuremberg," Wikipedia. en.wikipedia.org/wiki/1561_celestial_phenomenon_over_Nuremberg, accessed 9 Mar 2021.

"The Alaska Triangle: Unexplained Disappearances." YouTube. Uploaded by Top Mysteries, 17 Jan 2020, www.youtube.com/watch?v=nLb6tMDqiJw, accessed 9 Mar 2021.

Allen, Lauretta. Personal interview. 6 February 2021.

"Are the Mysterious Dogu Figurines Depictions of Alien Astronauts?" Ancient-Code.com, no upload date given, www.ancient-code.com/are-the-mysterious-dogu-figurines-depictions-of-ancient-astronauts/, accessed 12 Apr 2021.

Aranburu, Arantza, Juan Luis Arsuaga and Nohemi Sala. "The stratigraphy of the Sima de los Huesos (Atapuerca, Spain) and implications for the origin of the fossil hominin accumulation." Quaternary International 433 (2017) 5-21.

"Are our brains to blame when we 'see' ghosts and have religious visions?" bbc.co.uk, uploaded 31 May 2016, www.bbc.co.uk/bbcthree/article/bf88do1a-844e-460e-b807-bae9fe2b2ddf accessed 4 December 2021.

Arkowitz, Hal and Scott O. Lilienfield. "Why Science Tells Us Not to Rely on Eyewitness Accounts." Scientific American. Uploaded 1 Jan 2010. www.scientificamerican.com/article/do-the-eyes-have-it/, accessed 11 Mar 2021.

Barrett, Lisa Feldman and Daniel J. Barrett. "What Do Ghosts Feel?" Association for Psychological Science. Uploaded 30 September 2016. www.psychologicalscience.org/observer/what-do-ghosts-feel, accessed 3 September 2011.

Barttholomew, Robert E. and Joe Nickell. "The Amityville Hoax at 40: Why the Myth Endures." Skeptic Magazine, vol. 21, no. 4, 2016, pp. 8-12.

Berry, Richard B., MD. "Parasomnias." Science Direct. 2012. www.sciencedirect.com/topics/medicine-and-dentistry/hypnopompic, accessed 16 Apr 2021.

Billings, Lee. "Astronomer Avi Loeb Says Aliens Have Visited, and He's Not Kidding." Scientific American, 1 Feb 2021, www.scientificamerican.com/article/astronomer-avi-loeb-says-aliens-have-visited-and-hes-not-kidding1/, accessed 17 Mar 2021.

Bowers, Kenneth S., and John D. Eastwood. "On the Edge of Science: Coping With UFOlogy Scientifically." *Psychological Inquiry*, vol. 7, no. 2, Apr. 1996, p. 136.

Boyle, Rebecca. "The Milky Way is Disappearing." theatlantic.com, uploaded 10 Jun 2016, www.theatlantic.com/science/archive/2016/06/pawnee-sky/486557/, accessed 21 Mar 2021.

"The Bridgewater Triangle." Amazon Prime. www.amazon.com/Bridgewater-Triangle-Loren-Coleman/dp/B01LXMVMI, Accessed 6 Mar 2021.

Brookesmith, Peter. *UFO: The Complete Sightings*. New York: Barnes & Noble, 1995.

Campobasso, Craig. *The Extraterrestrial Species Almanac: The Ultimate Guide to Greys, Reptilians, Hybrids, and Nordics*. Massachusetts: Red Wheel, 2021.

Carbone, Jaclyn. "Near-death experience: London man shares what it's like to die – and then come back." Global news. Uploaded 5 February 2020, globalnews.ca/news/6511832/near-death-experience-london-man-shares-what-its-like-to-die-and-then-come-back/, accessed 21 November 2021.

Carbonell, Eudaid, Marina Mosquera, C.R. Palevol. "The emergence of a symbolic behaviour; the sephulchral pit of Sima de los Huesos, Sierra de Atapuerca, Burgos, Spain." Human Paleontology and Prehistory, 5 (2006) 155-160.

"The Celestial Bodies to Blame for Many UFO Sightings." YouTube. Uploaded by Smithsonian Channel, 4 Dec 2015, www.youtube.com/watch?v=KGawp7wBogk, accessed 9 Mar 2021.

Chabris, Christopher and Daniel Simons. "The Invisible Gorilla." theinvisiblegorilla.com, uploaded 2010, www.theinvisiblegorilla.com/gorilla_experiment.html, accessed 11 Mar 2021.

Chew, Stephen L. "Myth: Eyewitness Testimony is the Best Kind of Evidence." *Association for Psychological Science*, uploaded 20 Aug 2018, www.psychologicalscience.org/teaching/myth-eyewitness-testimony-is-the-best-kind-of-evidence.html, accessed 11 Mar 2021.

Choi, Charles Q. "'Alien Megastructure Star May Not Be So Special After All." Space.com, uploaded 19 Sep 2019. www.space.com/alien-megastructure-mysteriously-dimming-stars.html, accessed 16 Mar 2021.

Clancy, Susan A. *Abducted: How People Come to Believe They Were Kidnapped by Aliens*. Massachusetts: Harvard University Press, 2005.

Clark, Lindsay D. "Confrontation with Death Illuminagtes Death's Mystery in the *Odyssey*." Inquiries, Vol. 1 No. 11, 2009, Pg 1/1.

Clark, Steven E., and Elizabeth F. Loftus. "The Construction of Space Alien Abduction Memories." Psychological Inquiry, vol. 7, no. 2, Apr. 1996, p. 140.

Clarke, Ardy Sixkiller. *Space Age Indians: Their Encounters with the Blue Men, Reptilians, and Other Star People*. Texas: Anomalist Books, 2019.

Clarke, David. "Radar Angels." *Fortean Times 195* (2005), no upload date given, drdavidclarke.co.uk/radar-uaps/radar-angels/, accessed 17 Mar 2021.

Cookney, Francesca. "AMMACH: Britain's weirdest support group says 1,500 people are abducted by aliens each year." Mirror, 9 June 2013, www.mirror.co.uk/news/weird-news/ammach-britains-weirdest-support-group-194040, accessed 28 Apr 2021.

Cromie, William J. "Alien abduction claims examined: signs of trauma found." The Harvard Gazette, uploaded 20 Feb 2003, news.harvard.edu/gazette/story/2003/02/alien-abduction-claims-examined-2/, accessed 5 April 2021.

Curry, Eugene A. "The Final (Missions) Frontier: Extraterrestrials, Evangelism, and the Wide Circle of Human Empathy." *Zygon: Journal of Religion & Science*, vol. 54, no. 3, Sept. 2019, pp. 588–601.

Dagnall, Neil and Ken Drinkwater. "The top three scientific explanations for ghost sightings." The Conversation, uploaded 27 October 2016, theconversation.com/the-top-three-scientific-explanations-for-ghost-sightings-58259, accessed 3 September 2021.

Daugherty, Greg. "Meet J. Allen Hynek, the Astronomer Who First Classified UFO 'Close Encounters.'" History.com Updated 15 Jan 2020. www.history.com/news/j-allen-hynek-ufos-project-blue-book, accessed 10 Mar 2021.

Daugherty, Greg and Missy Sullivan. "Huge, Hovering and Silent: The Mystery of 'Black Triangle' UFOs." history.com, 22 Jul 2020, www.history.com/news/black-triangle-ufos-facts, accessed 18, Mar 2021.

Davisson, Zack. "Yurei: the Ghosts of Japan," Electric Lit, uploaded 23 October 2015. electricliterature.com/yurei-the-ghosts-of-japan/, accessed 4 December 2021.

DeGrazier, Michael, director. "Missing 411: The Hunted." Amazon Prime. 2019. www.amazon.com/gp/video/detail/B08B3CNH4C/ref=atv_dl_rdr?autoplay=1, accessed 14 Apr 2021.

Dewan, William J. "'A Saucerful of Secrets': An Interdisciplinary Analysis of UFO Experiences." *Journal of American Folklore,* vol. 119, no. 472, Spring 2006, pp. 184–202.

Dolan, Richard. *The Alien Agendas: A Speculative Analysis of Those Visiting Earth.* New York: Richard Dolan Press, 2020.

Draper, Scott, and Joseph O. Baker. "Angelic Belief as American Folk Religion." *Sociological Forum,* vol. 26, no. 3, Sept. 2011, pp. 623–643

Eby, Sharon. *Bigfoot Beyond Belief: A Study in Cultural Anthropology of What People Believe About Bigfoot/Sasquatch.* Nova Scotia: Sharon Eby, 2021.

Eghigian, Greg. "How UFO Reports Change with the Technology of the Times." Smithsonianmag.com, uploaded 1 Feb 2018, www.smithsonianmag.com/history/how-ufo-reports-change-with-technology-times-180968011/, accessed 18 Mar 2021.

Emerson. "You Won't Want to Visit This Notorious Nevada Cemetery Alone or After Dark," Only In Your State, uploaded 16 Feb 2019. www.onlyinyourstate.com/nevada/creepy-cemetery-nv/, accessed 4 December 2021.

Emmons, Charles. *Chinese Ghosts: Revisited.* Hong Kong: Blacksmith Books, 2017.

"Extraterrestrial Highway." TravelNevada. travelnevada.com/road-trip/extraterrestrial-highway/, accessed 5 Mar 2021.

"Europa Clipper." NASA.gov, no upload date given. europa.nasa.gov/europa/life-ingredients/, accessed 17 Mar 2021.

Farooq, Ayesha and Ashraf K. Kayani. "Prevalence of Superstitions and other Supernaturals in Rural Punjab: A Sociological Perspective." *South Asian Studies,* vol. 27, no. 2, July-December 2012, pp. 335-344.

Finkelstein, Joshua D. "The Ψ-Files: A Review of the Psychological Literature Regarding False Memories of Alien Abduction." *New School Psychology Bulletin,* vol. 14, no. 1, Jan. 2017, pp. 37–44.

Fitzpatrick, Molly. "Quarantining With a Ghost? It's Scary." The New York Times, uploaded 14 May 2020, nytimes.com/2020/05/14/style/haunted-house-ghost-quarantine.html, accessed 3 September 2021.

Freed, Ruth S. and Stanley A. Freed. "Ghost Illness of Children in North India." Med Anthropol. 1990 Nov;12(4):401-17.

Garber, Megan. "The Man Who Introduced the World to Flying Sarucers: Kenneth Arnold saw something, said something, and ushered in the UFO-industrial complex." The Atlantic. 15 Jun 2014. www.theatlantic.com/technology/archive/2014/06/the-man-who-introduced-the-world-to-flying-saucers/372732/, accessed 9 Mar 2021.

Garland, Robert. *The Greek Way of Death*. Ithaca, New York: Cornell University Press, 1985.

Gault, Matthew. "Researchers Think They Solved the Mystery of America's 'Lost Colony.'" Vice. 18 Aug 2020, www.vice.com/en/article/4aypdq/lost-colony-of-roanoke-mystery-solved-new-book-claims, accessed 30 Mar 2021.

Geiger, John. *The Third Man Factor: Surviving the Impossible*. New York: Weinstein Books, 2009.

Gentile, A.J. "The Nevada Triangle | 2,000 Planes Mysteriously Crashed & Missing Near Area 51." YouTube. Uploaded by The Why Files, 21 January 2021, www.youtube.com/watch?v=WMETBHvo-U4

Gibson, Arvin S. "Commenting on "Frightening Near-Death Experiences."" *Journal of Near-Death Studies*. 15(2), Winter 1996, 141-148.

Gilhus, Ingvile Sælid, Alexandros Tsakos, Marta Camilla Wright, eds. *The Archangel Michael in Africa: History, Cult, and Persona*. London: Bloomsbury Academic, 2021.

Godawa, Brian. *When Giants Were Upon the Earth: The Watchers, the Nephilim, and the Biblical Cosmic War of the Seed*. New York: Warrior Poet Publishing, 2021.

Grande, Todd. "Alien Abduction Story Analysis | Travis Walton/Joe Rogan Interview." YouTube. Dr. Todd Grande, uploaded 4 Feb 2021, www.youtube.com/watch?v=NB0RA6dnYKs, accessed 7 Apr 2021.

Green, Joseph P., et al. "Hypnosis and Psychotherapy: From Mesmer to Mindfulness." *Psychology of Consciousness: Theory, Research, and Practice*, vol. 1, no. 2, June 2014, pp. 199–212.

Green, Joseph P., et al. "Hypnotic Pseudomemories, Prehypnotic Warnings, and the Malleability of Suggested Memories." *Applied Cognitive Psychology*, vol. 12, no. 5, Oct. 1998, pp. 431–444.

Grush, Loren. "NASA is updating its guidelines on how to prevent contamination of the Solar System," theverge, uploaded 9 Jul 2020, www.theverge.com/2020/7/9/21318986/nasa-planetary-protection-guidelines-moon-mars-artemis-human-exploration, accessed 17 Mar 2021.

Habib, Ayesha. "The Call of the Voice: Why do we feel the urge to jump from high places?" nuvomagazine.com, uploaded 10 September 2021, nuvomagazine.com/magazine/autumn-2021/the-call-of-the-void, accessed 7 December 2021.

Hall, William J. and Jimmy Petonito. *Phantom Messages: Chilling Phone Calls, Letters, Emails, and Texts from Unknown Callers*. Massachusetts: Disinformation Books, 2018.

Harpur, Patrick. *Daimonic Reality: A Field Guide to the Otherworld*. Washington: Pine Winds Press, 1994.

Heaney, Christopher. "The Racism Behind Alien Mummy Hoaxes: Pre-Columbian bodies are once again being usd as evidence for extraterrestrial life." The Atlantic. Uploaded 1 Aug 2017, www.theatlantic.com/science/archive/2017/08/how-to-fake-an-alien-mummy/535251/, accessed 12 Apr 2021.

"Hearing ghost voices relies on pseudoscience and fallibility of human perception," the conversation, uploaded 30 October 2015. theconversation.com/hearing-ghost-voices-relies-on-pseudoscience-and-fallibility-of-human-perception-48160, accessed 4 December 2021.

Hendrickson, Rachel. "The Virginia City Great Fire of 1875." Intermountain Histories, uploaded 29 May 2019. https://www.intermountainhistories.org/items/show/246, accessed 4 December 2021.

Hollars, B. J. "In Defense of Sasquatch." *Ninth Letter*, vol. 8, no. 1, Spring/Summer 2011, pp. 59–66.

Homo heidelbergensis. Smithsonian National Museum of Natural History. Uploaded 22n January 2021. humanorigins.si.edu/evidence/human-fossils/species/homo-heidelbergensis, accessed 13 September 2021.

Horselenberg, Robert, et al. "Individual Differences in the Accuracy of Autobiographical Memory." *Clinical Psychology & Psychotherapy*, vol. 11, no. 3, May 2004, pp. 168–176.

Houran, James and Rense Lange. "Tolerance of Ambiguity and Fear of the Paranormal in Self-Identified Percipients of Haunting/RSPK Phenomena." *Journal of the Society for Psychical Research*, vol. 62, no. 848, July 1997, pp. 36-40.

"How Color Affects Taste: A Lesson in Gastrophysics." Food Republic, uploaded 29 Jun 2017. medium.com/@foodrepublic/how-color-affects-taste-a-lesson-in-gastrophysics-1c1d3df89702, accessed 12 Mar 2021.

Hulce, Kyle. "The 13 Most Terrifying Ghosts in Thailand," the culture trip, uploaded 13 August 2018, theculturetrip.com/asia/thailand/articles/13-terrifying-ghosts-thai-folklore/, accessed 4 December 2021.

"I Was Abducted by Aliens." YouTube. Uploaded by Truly, 17 Jan 2021, www.youtube.com/watch?v=IU6UPMTKazY, accessed 31 Jan 2021.

"Indrid Cold – Casefiles #1." YouTube. Small Town Monsters, uploaded 1 Nov 2017. www.youtube.com/watch?v=_0JfU3ch-AY, accessed 22 Mar 2021.

Irwin, Neil. "For New UFO Lobby, 'X-Files' Are Real." Christian Science Monitor, vol. 91, no. 183, 17 Aug. 1999, p. 5. EBSCOhost, search.ebscohost.com/login.aspx?direct=true&db=aph&AN=2159350&site=ehost-live&scope=site.

Janssen, Volker. "How the 'Little Green Men' Phenomenon Began on a Kentucky Farm." History.com, uploaded 2 Jan 2020. www.history.com/news/little-green-men-origins-aliens-hopkinsville, accessed 6 Apr 2021.

Jiang, Fercility. "Hungry Ghost Festival." China Highlights, uploaded 2 October 2021. www.chinahighlights.com/festivals/hungry-ghost-festival.htm, accessed 4 December 2021.

Johanson, Donald and Maitland Edey. *Lucy: The Beginnings of Humankind.* New York: Warner Books, 1981.

Josh@asia_backpackers. "Ghost Spirits and Thai Folklore," thailand discovery, uploaded 4 March 2016, www.thailanddiscovery.info/ghosts-spirits-and-thai-folklore/, accessed 5 December 2021.

Jung, C. G. "A Visionary Rumour." Journal of Analytical Psychology, vol. 4, no. 1, Jan. 1959, pp. 5–19.

Keel, John A. *The Complete Guide to Mysterious Beings.* New York: Doubleday, 1970.

Kelley-Romano, Stephanie. "Mythmaking in Alien Abduction Narratives." *Communication Quarterly*, vol. 54, no. 3, Aug. 2006, pp. 383–406.

Kelly, Meghan B., "The Science Behind the 'Call of the Void,'" wbur.com, uploaded 29 June 2018, wbur.org/endlessthread/2018/06/29/the-call-of-the-void, accessed 7 December 2021.

Kiessling, Nicolas K. "Grendel: A New Aspect." *Modern Philology*, vol. 65, no. 3, Feb 1968, pp. 191-201.

Klangboonkrong, Manta. There's more to a Thai ghost story than being scary," thailandnow.in.th, no upload date given, www.thailandnow.in.th/arts-culture/theres-more-to-a-thai-ghost-story-than-being-scary/, accessed 4 December 2021.

Kloor, Keith. "The Media Loves This UFO Expert Who Says He Worked for an Obscure Pentagon Program. Did he?" The Intercept. Uploaded 1 Jun 2019, theintercept.com/2019/06/01/ufo-unidentified-history-channel-luis-elizondo-pentagon/, accessed 25 Mar 2021.

Kloor, Keith. "UFOs Won't Go Away." *Issues in Science & Technology*, vol. 35, no. 3, Spring 2019, pp. 49–56.

Koch, Christof. "What Near-Death Experiences Reveal about the Brain." *Scientific American* Uploaded 1 June, 2020. www.scientificamerican.com/article/what-near-death-experiences-reveal-about-the-brain/, accessed 21 November 2021.

Lamberg, Lynne. "Belief in Alien UFOs Deep in American Psyche." *JAMA*, vol. 278, no. 3, July 16, 1997, pp. 193.

Landau, Elizabeth. "What We Know—And Don't Know—About 'Oumuamua." NASA, 27 Jun 2018. solarsystem.nasa.gov/news/473/what-we-knowand-dont-knowabout-oumuamua/, accessed 17 Mar 2021.

Lange, Rense and James Houran. "Context-Induced Paranormal Experiences Support for Houran and Lange's Model of Haunting Phenomena." *Perceptual and Motor Skills*, vol. 94, no. 3, 1997, pp. 1455-1458.

"Life on Titan." European Space Agency, no upload date given, esa.int/Science_Exploration/Space_Science/Cassini-Huygens/Life_on_Titan, accessed 17 Mar 2021.

Lumpkin, Joseph B. *The Books of Enoch: The Angels, The Watchers, and the Nephilim.* Alabama: Fifth Estate Publishers, 2015.

Lynn, Heather. *The Anunnaki Connection: Sumerian Gods, Alien DNA & The Fate of Humanity.* Massachusetts: New Page Books, 2020.

Lynn, Heather. *Evil Archaeology: Demons, Possessions, and Sinister Relics.* Massachusetts: Disinformation Books, 2019.

Lyon, Jason. "To Pay Attention, the Brain Uses Filters, Not a Spotlight." Quanta Magazine. 24 Sep 2019. www.quantamagazine.org/to-pay-attention-the-brain-uses-filters-not-a-spotlight-20190924/, accessed 11 Mar 2021.

Magis Center. "5 Credible Near Death Experience (Peer-Reviewed)," magiscenter.com, uploaded 19 Aug 2020. blog.magiscenter.com/blog/credible-near-death-experience-stories, accessed 4 December 2021.

Markovsky, Barry. "Notes on a Haunting: How Science Can Explain Ghosts and Haunted Houses." *Skeptic Magazine,* vol. 26, no. 2, 2021, pp. 37-45.

Martin, Jean-Rémy, and Elisabeth Pacherie. "Alterations of Agency in Hypnosis: A New Predictive Coding Model." *Psychological Review*, vol. 126, no. 1, Jan. 2019, pp. 133–152.

"The Maury Island Incident." HowStuffWorks, Uploaded by the Editors of Publications International, LTD. science.howstuffworks.com/space/aliens-ufos/maury-island-incident.htm, accessed 9 Mar 2021.

McClenon, James. "A Community Survey of Psychological Symptoms: Evaluating Evolutionary Theories Regarding Shamanism and Schizophrenia." Mental Health, Religion & Culture, vol. 15, no. 8, Oct. 2012, pp. 799–816.

McGaha, James and Joe Nickell. "The Roswellian Syndrome: How Some UFO Myths Develop." Skeptical Inquirer, Volume 36, No. 3, May/June 2012. skepticalinquirer.org/2012/05/the-roswellian-syndrome-how-some-ufo-myths-develop/, accessed 9 Mar 2021.

"McGurk Effect – Auditory Illusion – BBC Horizon Clip." YouTube.
  Uploaded by sixfullofnines, 16 Mar 2016,
  www.youtube.com/watch?v=2k8fHR9jKVM
McLachlan, Sean. *Hollow Earth: A History of the Strange Tales, Bizarre Beliefs,
  and Conspiracy Theories about the Earth's Core.* Illinois: Charles River
  Editors, 2017.
McRobbie, Linda Rodriguez. "Why alien abductions are down dramatically."
  Boston Globe. Uploaded 12 Jul 2016.
  www.bostonglobe.com/ideas/2016/06/11/why-alien-abductions-are-down-
  dramatically/qQ3zdBIc2tLAf3LVms8GLP/story.html, accessesd 14 Apr
  2021.
Meldrum, Jeff. "Sasquatch & Other Wildmen: The Search for Relict
  Hominoids." *Journal of Scientific Exploration,* vol. 30, no. 3, Fall 2016, pp.
  355–373.
Mencken, F.Carson, et al. "Round Trip to Hell in a Flying Saucer: The
  Relationship between Conventional Christian and Paranormal Beliefs in the
  United States." *Sociology of Religion,* vol. 70, no. 1, Spring 2009, pp. 65–85.
"Missing 411—Behind the Mysteries: Strange disappearances in national parks."
  Paranormal Authority, no upload date given,
  paranormalauthority.com/missing-411/, accessed 30 Mar 2021.
Mobley, Gregory. "The Wild Man in the Bible and the Ancient Near East."
  Journal of Biblical Literature, vol. 116, no. 2, Summer 1997, p. 217.
Morris, Gregory L. "Imagining Bigfoot." *Western American Literature,* vol. 42,
  no. 3, Fall 2007, pp. 276–292.
"The Most Horrifying Ghosts in Thailand," thailand insider, uploaded 29
  October 2020, thailandinsider.com/the-most-horrifying-ghosts-in-thailand/,
  accessed 4 December 2021.
Nees, Michael. "Hearing ghost voices relies on pseudoscience and fallibility of
  human perception." *The Conversation,* Uploaded 30 October 2015.
  Theconversation.com/hearing-ghost-voices-relies-on-pseudoscience-and-
  fallibility-of-human-perception-48160, accessed 15 November 2021.
Newman, Leonard S., and Roy F. Baumeister. "Toward an Explanation of the
  UFO Abduction Phenomenon: Hypnotic Elaboration, Extraterrestrial
  Sadomasochism, and Spurious Memories." *Psychological Inquiry,* vol. 7, no.
  2, Apr. 1996, p. 99
Nickell, Joe. "Famous Alien Abduction In Pascagoula: Reinvestigating a Cold
  Case." Skeptical Inquirer Volume 36, No. 3.
  skepticalinquirer.org/2012/05/famous-alien-abduction-in-pascagoula-
  reinvestigating-a-cold-case/, accessed 9 Mar 2021.

Nickell, Joe, Barry Karr and Tom Genomi, editors. *The Outer Edge: Classic Investigations of the Paranormal.* New York: CSICOP, 1996.

Nie, Fanhao, and Daniel V. A. Olson. "Demonic Influence: The Negative Mental Health Effects of Belief in Demons." *Journal for the Scientific Study of Religion*, vol. 55, no. 3, Sept. 2016, pp. 498–515.

Nita, Maria. "Sky vs. Earthly Empowerment: From Angels and Superheroes to Humans and Community in the Marvel Universe and Green Christian Cosmology." *Journal of Religion & Popular Culture*, vol. 31, no. 3, Sept. 2019, pp. 236–249.

Novotney, Amy. "The risks of social isolation." apa.org, uploaded 15 May 2019, Vol 50, No. 5. apa.org/monitor/2019/05/ce-corner-isolation, accessed 17 Mar 2021.

Oberding, Janice. *Ghosts of Goldfield and Tonopah.* Charleston, SC: Haunted America, 2015.

Offutt, Jason. *Darkness Walks: The Shadow People Among Us.* New York: Anomalist Books, 2009.

"Old Tonopah Cemetery Walking Tour." Tonopah Nevada, no upload date given. www.tonopahnevada.com/old-tonopah-cemetery/, accessed 4 December 2021.

Orne, Martin T. and A. Gordon Hammer, eds. "Hypnosis." Encyclopaedia Britannica. No upload date given. www.britannica.com/science/hypnosis, accessed 19 Apr 2021.

Patry, Alain L., and Luc G. Pelletier. "Extraterrestrial Beliefs and Experiences: An Application of the Theory of Reasoned Action." *Journal of Social Psychology*, vol. 141, no. 2, Apr. 2001, pp. 199–217.

Penascal, Maria. "Yurei: The Japanese Culture of Ghosts through History," voyapon.com, uploaded 13 October 2020, voyapon.com/yurei-japanese-ghosts, accessed 4 December 2021.

Petrich, Loren. "Close Encounters of the Various Kinds." lpetrich.org/UFOs/Close%20Encounters.xhtml, accessed 9 Mar 2021.

Phelan, Matthew. "Navy Pilot Who Filmed the 'Tic Tac' UFO Speaks: 'It Wasn't Behaving by the Normal Laws of Physics.'" New York Intelligencer, 19 Dec 2019, nymag.com/intelligencer/2019/12/tic-tac-ufo-video-q-and-a-with-navy-pilot-chad-underwood.html, accessed 18 Mar 2021.

Piper, Grant. "The Earliest Ghost Stories." Medium.com, 31 October 2020, medium.com/exploring-history/the-earliest-ghost-stories-93643d893203, accessed 5 October 2021.

Poppy, Carrie. "A scientific approach to the paranormal," Ted.com, uploaded October 2016,

www.ted.com/talks/carrie_poppy_a_scientific_approach_to_the_paranormal
/transcript?language=en, accessed 4 December 2021.

"Problems With Witness Testimony: Tricks Memory Plays."
mcadams.posc.mu.edu, No Upload Date Given.
mcadams.posc.mu.edu/memory.htm, accessed 11 Mar 2021.

Prothero, Donald. "The Hollow Earth: If You Thought the Flat Earthers Were
Out There, Wait Until You Read About Those Who Think the Earth Is an
Empty Sphere Filled With Wonders." *Skeptic*, vol. 25, no. 3, July 2020, pp.
18–23.

Pugliese, David. "Journey to Area 51: Black helicopters and claims of abductions
by aliens." Ottawa Citizen, uploaded 19 Sept 2019,
ottawacitizen.com/news/national/defence-watch/journey-to-area-51-black-
helicopters-and-claims-of-alien-abductions, accessed 22 Mar 2021.

Putsch III, Robert W. "Ghost Illness: A Cross-Cultural Experience With the
Expression of a Non-Western Tradition in Clinical Practice." *American
Indian and Alaska Native Mental Health Research*, 2(2), pp. 6-26.

Pyle, Robert Michael. *Where Bigfoot Walks: Crossing the Dark Divide*.
Berkeley, California: Counterpoint, 1995.

Radford, Benjamin. "The Shady Science of Ghost Hunting." Livescience.com,
uploaded 27 October 2006, www.livescience.com/4261-shady-science-
ghost-hunting.html, accessed 4 December 2021.

Randle, Kevin D. *The Government UFO Files: The Conspiracy of Cover-Up*.
Michigan: Visible Ink Press, 2014.

"Recent Adverse Publicity on Parapsychological Research." CIA.gov.
www.cia.gov/readingroom/docs/CIA-RDP96-00788R001100360001-
1.pdf, accessed 11 Mar 2021.

Ricksecker, Mike. *A Walk in the Shadows: A Complete Guide to Shadow People*,
2nd edition. USA; Haunted Road Media, 2021.

Rittichainuwat, Bongkosh. "Ghosts: A travel barrier to tourism
recovery." *Annals of Tourism Research*, vol. 38,2 (2011): 437-459.
doi:10.1016/j.annals.2010.10.001

Robin, Frédérique, et al. "Hypnosis and False Memories." *Psychology of
Consciousness: Theory, Research, and Practice*, vol. 5, no. 4, Dec. 2018, pp.
358–373.

Robson, David. "Psychology: The truth about the paranormal." BBC Future,
uploaded 30 October 2014, www.bbc.com/future/article/20141030-the-
truth-about-the-paranormal, accessed 3 September 2021.

Rogo, D. Scott. *The Haunted House Handbook*. Tempo Books: United States,
1978.

Rojas, Alejandro. "New survey shows nearly half of Americans believe in aliens." Huffpost, Uploaded 02 Aug 2017, www.huffpost.com/entry/new-survey-shows-nearly-half-of-americans-believe-in_b_59824c11e4b03d0624b0abe4, accessed 29 Mar 2021.

Rojcewicz, Peter M. "The 'Men in Black' Experience and Tradition: Analogues with the Traditional Devil Hypothesis." The Journal of American Folklore, vol. 100, no. 396, 1987, pp. 148–160. JSTOR, www.jstor.org/stable/540919. Accessed 22 Mar. 2021.

Roos, Dave. "When UFOs Buzzed the White House and the Air Force Blamed the Weather." History. www.history.com/news/ufos-washington-white-house-air-force-coverup, accessed 9 Mar 2021.

Rose, Steve. "The real Men in Black, Hollywood and the great UFO cover-up." theguardian.com, uploaded 14 Aug 2014. www.theguardian.com/film/2014/aug/14/men-in-black-ufo-sightings-mirage-makers-movie, accessed 22 Mar 2021.

Sayers, William. "Middle English 'Wodewose' 'Wilderness Being': A Hybrid Etymology?" ANQ, vol. 17, no. 3, Summer 2004, pp. 12–20.

Seemangal, Robin. "Meet the Lobbyist Pressuring the US Gov't to Disclose Extraterrestrial Activity." Observer, uploaded 16 Sep 2015, observer.com/2015/09/meet-the-lobbyist-pressuring-the-us-government-to-disclose-extraterrestrial-activity/, accessed 23 Mar 2021.

Segura, Olga. "True alienation: when a person of color tries to fit in with UFO enthusiasts." theguardian.com, uploaded 6 Mar 2020, www.theguardian.com/world/2020/mar/06/aliens-ufos-olga-segura, accessed 24 Mar 2021.

Seth, Anil. "Your Brain Hallucinates Your Conscious Reality." YouTube. Uploaded by TED, 18 Jul 2017, www.youtube.com/watch?v=lyu7v7nWzfo, Accessed 7 Mar 2021.

Sheaffer, Robert. The UFO Verdict: Examining the Evidence. New York: Prometheus Books, 1986.

Shinn, Sharon. Personal Correspondence, 15 Apr 2021.

Shubin, Neil. Your Inner Fish: A Journey Into the 3.5-Billion-Year History of the Human Body. New York: Vintage Books, 2009.

Siegel, Ethan. "The 5 Possibilities for Life on Mars." Forbes, Uploaded 4 Aug 2020, www.forbes.com/sites/startswithabang/2020/08/04/the-5-possibilities-for-life-on-mars/?sh=73c24d195387, accessed 17 Mar 2021.

Simon, Edward. "Why Sasquatch and Other Crypto-Beasts Haunt Our Imaginations." Anthropology of Consciousness, vol. 28, no. 2, Fall 2017, pp. 117–120.

"Solar System Exploration: Titan," NASA.gov, no upload date given, solarsystem.nasa.gov/moons/saturn-moons/titan/in-depth/, accessed 17 Mar 2021.

Spanos, Nicholas P., and Patricia A. Cross. "Close Encounters: An Examination of UFO Experiences." *Journal of Abnormal Psychology*, vol. 102, no. 4, Nov. 1993, p. 624.

Spitznagel, Eric. "Why hundreds of people vanish into the American wilderness." New York Post, uploaded 4 Jul 2020, nypost.com/2020/07/04/why-hundreds-of-people-vanish-into-the-american-wilderness/, accessed 29 Mar 2021.

Steffen, Andrea D. "Researchers Have Made Self-Assembling DNA Nanobots with Encoded Structural Plans," Intelligent Living, Uploaded 18 Feb 2021, www.intelligentliving.co/self-assembling-dna-nanobots/, accessed 26 Apr 2021.

"Step Back in Time Virginia City, Nevada, Est. 1859." Virginia City NV, no upload date given. https://visitvirginiacitynv.com/history/, accessed 4 December 2021.

Stephey, M.J. "A Brief History of UFOs." *Time.* 17 Dec 2009. content.time.com/time/health/article/0,8599,1948214,00.html. Accessed 5 Mar 2021.

Stockton, Steve. *Strange Things In the Woods: A Collection of Terrifying Stories.* Beyond the Fray Publishing, 2013.

Stover, Dawn. "Double Dread: UFOs and Nuclear War." the bulletin.org, 4 Jun 2019, thebulletin.org/2019/06/double-dread-ufos-and-nuclear-war/, accessed 20 Mar 2021.

Sumner, Mark. Personal Correspondence, 9 Apr 2021.

Swami, Viren, et al. "The Truth Is Out There: The Structure of Beliefs About Extraterrestrial Life Among Austrian and British Respondents." *Journal of Social Psychology,* vol. 149, no. 1, Feb. 2009, pp. 29–43.

Swami, Viren, et al. "Psychology in Outerspace: Personality, Individual Difference, and Demographic Predictors of Beliefs about Extraterrestrial Life." *European Psychologist,* vol. 15, no. 3, 2010, pp. 220–228.

Szanto, Katalin, et al. "Indirect self-destructive behavior and overt suicidality in patients with complicated grief," *Journal of Clinical Psychiatry,* 2006 Feb;67(2):233-9.

Takyi, Daniel. "Why we actually want to jump: The Call of the Void explained," the Bubble.org.uk, uploaded 10 December 2019, thebubble.org.uk/current-affairs/science-technology/whh-we-actually-want-to-jump, accessed 7 December 2021.

Taylor, Timothy. *The Buried Soul: How Humans Invented Death.* Boston: Beacon Press, 2002.

Taylor, Troy. *The Ghost Hunter's Guidebook.* Alton, IL: Whitechapel Productions Press, 2001.

"Terror in the Skies." Amazon Prime, www.amazon.com/Terror-Skies-Thunderbirds-Prehistoric-Remnants/dp/B07RNN1NFK, accessed 6 Mar 2021.

"Thirteen Haunted Spots in Virginia City." RenoTahoe, www.visitrenotahoe.com/plan-your-trip/region/haunted-places-in-virginia-city/, accessed 4 December 2021.

"UFO Hunters: Full Episode-Reverse Engineering (Season 1, Episode 7) | History." YouTube. History.com. 24 Feb 2019, www.youtube.com/watch?v=GokNWVlpius, accessed 11 Mar 2021.

"UFO Hunters: Terrifying Encounters with Mysterious Beings (S3, E12) | Full Episode | History." YouTube. History.com. 8 Mar 2021, www.youtube.com/watch?v=w7z7u6enuNw&t=39s, accessed 9 Mar 2021.

"UFO Sightings Surge Across US | Unidentified (Season 2) | History." YouTube. Uploaded by History, 4 Sep 2020, www.youtube.com/watch?v=y-6rgZwYo4g, accessed 9 Mar 2021.

"UFOstalker.com." Accessed 10 Mar 2021.

Unacknowledged. Mike Mazzola. Amazon Prime, 2017.

Vallee, Jacques. *Passport to Magonia: From Folklore to Flying Saucers.* Brisbane, Australia: Daily Grail Publishing, 1969.

Vidyasagar, Aparna. "What is CRISPR?" LiveScience. Uploaed 21 Apr 2018. www.livescience.com/58790-crispr-explained.html, accessed 12 Apr 2021.

Wagstaff, Graham F. "Is There a Future for Investigative Hypnosis?" *Journal of Investigative Psychology & Offender Profiling*, vol. 6, no. 1, Jan. 2009, pp. 43–57.

Wall, Mike. "UFOs Are Real, But Don't Assume They're Alien Spaceships," space.com, 31 May 2019, www.space.com/ufos-real-but-not-alien-spaceships.html, accessed 17 Mar 2021.

Webster, Donovan. "In 1947, A High-Altitude Balloon Crash Landed in Roswell. The Aliens Never Left: Despite its persistence in popular culture, extraterrestrial life owes more to the imagination than reality." *Smithsonian Magazine*, 5 Jul 2017, www.smithsonianmag.com/smithsonian-institution/in-1947-high-altitude-balloon-crash-landed-roswell-aliens-never-left-180963917/, accessed 9 Mar 2021.

Welfare, Simon and John Fairley. *Arthur C. Clarke's Mysterious World.* New York: Trident International Television Enterprises, 1980.

Welsh, Tim. "It feels instantaneous, but how long does it really take to think a thought?" *The Conversation*. 26 June 2015. www.theconversation.com/it-feels-instantaneous-but-how-long-does-it-really-take-to-think-a-thought-42392 Accessed 6 Mar 2021.

Wen, Tiffanie. "Why do people believe in ghosts?" The Atlantic. Uploaded 5 September 2014. www.theatlantic.com/health/archive/2014/09/why-do-people-believe-in-ghosts/379072/, accessed 3 Septemer 2021.

Westrum, Ron. "Social Intelligence about Anomalies: The Case of UFOs." Social Studies of Science, vol. 7, no. 3, 1977, pp. 271–302. JSTOR, www.jstor.org/stable/284599, accessed 5 Mar. 2021.

Westwood, Gail. "Summit Spirits: What is the difference between a ghost and a spirit?" Summit Spirits, uploaded 24 October 2015, www.summitdaily.com/news/summit-spirits-what-is-the-difference-between-a-ghost-and-a-spirit/, accessed 3 September 2021.

Whalen, Andrew. "Are Aliens Real? One-Third of Americans Think Alien UFOs Have Visited Earth." *Newsweek*, uploaded 6 Sep 2019, www.newsweek.com/aliens-are-real-ufos-2019-sightings-americans-area-51-raid-extraterrestrial-disclosure-1458103, accessed 16 Mar 2021.

Whalen, Andrew. "What if Aliens Met Racists? MUFON Resignations Highlight Internal Divisions in UFO Sightings Organization." *Newsweek*, uploaded 29 Apr 2018, www.newsweek.com/ufo-sightings-mufon-2018-john-ventre-alien-extraterrestrial-905060, accessed 24 Mar 2021.

"What is Nanotechnology?" nano.gov, no upload date given, www.nano.gov/nanotech-101/what/definition, accessed 18 Mar 2021.

"What is Sleep Paralysis?" Sleep Foundation. 6 Aug 2020. www.sleepfoundation.org/parasomnias/sleep-paralysis, accessed 16 Apr 2021.

"Where are memories stored in the brain?" Queensland Brain Institute. Uploaded 23 Jul 2018, qbi.uq.edu.au/brain-basics/memory/where-are-memories-stored#:~:text=The%20hippocampus%2C%20located%20in%20the,with%20a%20friend%20last%20week, accessed 19 Apr 2021.

Wilkins, Jacob. "The Nuremberg UFO Sighting of 1561," *Medium.com*, 18 Nov 2020, medium.com/lessons-from-history/the-nuremberg-ufo-sighting-of-1561-4078ecfcd946, accessed 9 Mar 2021.

Wiseman, Richard. *Paranormality: Why we see what isn't there*. London: MacMillan, 2011.

Witze, Alexandra. "Prospects for Life on Venus Fade—But Aren't Dead Yet." *Nature*, uploaded 17 Nov 2020. www.nature.com/articles/d41586-020-03258-5, accessed 17 Mar 2021.

Woody, Erik, and Pamela Sadler. "Interpersonal Aspects of Hypnosis: Twisted Pears and Other Forbidden Fruit." *Psychology of Consciousness: Theory, Research, and Practice*, Apr. 2020.

Zhang, Wenli. "How Do We Think About Death? A Cultural Glance of Superstitious Ideas from Chinese and Western Ghost Festivals." *International Education Studies*, vol. 2, no. 4, November 2009, pp. 68-71.

Index

About the Author

Marella Sands is a native St. Louisan who has published novels, novellas, short stories, a poem, an essay, and non-fiction works. Her historical novels, *Sky Knife* and *Serpent and Storm*, were set in 5th century Central America. In addition, she co-wrote two King's Quest novels with fellow St. Louisan Mark Sumner under the name Kenyon Morr. She has had short stories in several anthologies. She has a series set in an alternate United States, which is published by Ring of Fire Press. She also writes the Angels' Share books. She and her husband travel whenever they can. Marella earned degrees in anthropology from the University of Tulsa and Kent State University.

Word Posse Fun Fact

I've always loved stories of the unusual. Freak storms where frogs rained from the sky? UFOs over Ohio? Aliens and Men in Black wandering about the country? The Mothman of Point Pleasant? Ghosts in old castles? Strange shadows in the basement? Something big and hairy creeping through your campsite? I was hooked on it all. If it were weird, I was reading about it, whether it was Carl Sagan's *The Demon-Haunted World,* John Keel's *The Complete Guide to Mysterious Beings,* or Erich von Däniken's *Chariots of the Gods.* On television, I watched *Project U.F.O.* and *In Search Of.* Needless to say, when *The X-Files* landed on TV, I was there. While I didn't grow up to be a believer in the paranormal, I'm never *not* fascinated with the tales. I hope, with this series, to discuss possible explanations for various phenomena, as well as provide a forum for people to tell their stories. If you have a story you'd like to share, contact me at msands@marellasands.com. Or fill in my Google form at: https://tinyurl.com/eej3wt8. Or go to my website (www.writnfool.com) and click on the link on the right that says "My Paranormal Story." Thanks!

www.ingramcontent.com/pod-product-compliance
Lightning Source LLC
Chambersburg PA
CBHW050356280326
41933CB00010BA/1489